WEATHER
FLYING

Robert N. Buck

WEATHER FLYING

Revised Edition

Introduction by **Wolfgang Langewiesche**

DRAWINGS BY PIERRE BARRE

Macmillan Publishing Co., Inc.

·

New York

Collier Macmillan Publishers

·

London

Macmillan Publishing Co., Inc.
866 Third Avenue, New York, N.Y. 10022
Collier Macmillan Canada, Ltd.

Library of Congress Cataloging in Publication Data
Buck, Robert N   1914–
Weather flying.
Bibliography: p.
Includes index.
1.   Meteorology in aeronautics.   I.   Title.
TL556.B77   1978   629.132′5214   78–6513
ISBN 0–02–518020–7

First Revision 1978

Third Printing 1979

Designed by Jack Meserole

Printed in the United States of America

For Leighton Collins

CONTENTS

ABOUT SOME PEOPLE

THERE ARE many people to thank for helping me with this book.

It starts with a meteorologist named Mr. Ball, who, when I was sixteen and starting across the United States alone, briefed me at Newark Airport, New Jersey, on what the weather would be. I was interested in what he said because I didn't know much about weather, or flying either, and I was a little nervous.

Mr. Ball—I never knew his first name—also awakened my interest in weather, what makes it, and what a pilot does about it. Looking back, I realized that Mr. Ball was an exceptionally good meteorologist.

After Mr. Ball came a host of meteorologists, most of whose names I don't know and never will. They are voices and faces over a telephone or in a weather office in some city when I was trying to make cross-country records, or just going somewhere.

Some time later they were in wood-paneled rooms in Gander, Newfoundland, or Prestwick, Scotland, during World War II when the Atlantic, sitting out there waiting to be crossed, was a big, black unknown, and whatever that man said I listened to carefully because any crumb of information was something to grab for help in doing a job you didn't know all there was to know about doing.

There are meteorologists I got to know well by searching conversation as I tried to learn, and later in consultation as we probed the atmosphere in research—men like Ed Minser and J. A. Browne of TWA. I continued to learn from TWA meteorologists who briefed me on what the weather would be over the North Atlantic which became an ocean I respected, but felt very comfortable flying over in a 747. Now, retired, I occasionally call TWA's Met people if there's a sticky weather situation and I'm going to fly my airplane or glider.

There is Nieut Lieurance of the National Oceanic and Atmospheric Administration (NOAA), the new name for the Weather Bureau. Nieut is a pilot's meteorologist and has done much to bring good weather services to us all. Chuck Lindsay of NOAA is one of the world's foremost soaring meteorologists and a person always good for a new piece of knowledge. I thank Chuck also for reading this manuscript and making some valuable suggestions.

Lt. General Ira C. Eaker, a boyhood hero of mine, saw the possibilities of thunderstorm research and made it possible for TWA and the Air Force to work together using a P-61 Black Widow night fighter which I was fortunate enough to fly in the project.

Also, Dan Sowa, the superb meteorologist of Northwest Orient Airlines, has given me many insights into the mysteries of weather, as has his associate, the brilliant technical airman, Captain Paul A. Soderlind of Northwest Orient Airlines.

Years ago I learned from Dr. Horace Byers and for a brief but exciting and privileged time from Dr. Irving Langmuir.

I wish I could name all the others, but there isn't space. I do want it to be clear, however, that I thank all meteorologists. They are a misunderstood group who are cussed more than praised, when it should be the other way around.

I want to thank Velma and Dwane Wallace for urging me to get this book done.

Solid advice and encouragement came from Charles W. Ferguson and Caroline Rogers of *The Reader's Digest*. They are patient and understanding friends.

I must mention the man to whom this book is dedicated, Leighton Collins, who with his *Air Facts* magazine started me writing about weather almost forty years ago. His enthusiasm and encouragement did the most.

My wife Jean, daughter Ferris, and son Rob, an experienced and weatherwise airman himself, gave their enthusiastic support, and I thank them for that.

And very important is the lady who typed all this many times, but always willingly and efficiently, Hattie Fretz.

ROBERT N. BUCK

Fayston, Vermont

INTRODUCTION

WHAT CAPTAIN ROBERT BUCK says in *Weather Flying* has not been said before. Other books explain how weather is made; this book explains how weather is flown.

We get to the field in the morning. Here is the weather: the map, the forecasts, the sequence of current reports from many points, the winds aloft—the whole package, prepared by experts. However deep our knowledge of meteorology may be, we cannot hope to do better. For us as pilots, the question is: What do I *do* with this? Go or no go? If go: go underneath the weather, or on top, or through? Or go around it? Follow a railroad? File an instrument flight plan? Go right away? Delay a couple of hours?

Those questions. They are the last of the real problems of flying. Everything else has now been quite well mastered. The airplane itself now works: it handles nicely (at least, those flown by the general public). It has climb to spare, and we can usually find some level where the air is smooth. Noise and vibration are subdued: we can stand long hours in a day and make big distances. Electronics tells us where we are. Airports are plentiful and runways long. Engine failure is so rare we now almost forget about it. Even the economics of flying are no longer so forbidding. But the weather . . .

It is not often clearly impossible. When it is, we have no

problem: back to the hotel. On the contrary (as Captain Buck points out) the weather is normally flyable. Again, no problem, or only easy ones. But every once in a while—depending on season and part of the world—something is sitting out there that worries us. If you fly far enough in a straight line, you're likely to come up against some problem weather that very day. How well we deal with those situations determines how well the airplane works.

Too bold, and we cause emergencies and have accidents. Too timid, and we destroy the utility of the airplane and let our skill as pilots atrophy. Then pretty soon we have to be more timid still.

These weather decisions can be painful to make because we don't really know how to go about making them. And we know we don't know! We make them often by a process which is a dumb, confused struggle between "guts" and "judgment," ambition and fear. How it comes out depends on how we felt that day when we last had a good scare, whether the girlfriend is looking, or "This town is full, there is a convention on, you can't get a hotel room, let's go." Things like that tip the balance. And so (in Buck's words) we "drag the luggage back to town," often for no reasoned cause, or we "fling ourselves into the air," often with only a vague estimate of what's ahead. Or else, more likely, we stand around for another hour, look at the weather map some more, wait for the next weather sequence to come on the teletype and think, "I wish I could talk this over with some really experienced friend."

That's where Bob Buck comes in.

Experienced? We could consider 2000 hours quite a respectable lot of time. Buck has some 2000 Atlantic crossings! And those are only half of his experience. At this writing,

Captain Buck's log records well over 29,000 hours, all types. He set his first record when he was 16, in a Pitcairn Mailwing, an open-cockpit biplane. It was the Junior Transcontinental Record, New York to Los Angeles, and "It didn't amount to much—mostly just getting there." He set another record, for nonstop distance in light airplanes, in a 90-horsepower Monocoupe, overnight from California to Ohio—his engine quit there. With the same airplane he joined an expedition and searched the jungle of Yucatán for Maya ruins. He joined TWA as a co-pilot at the time when the DC2 was the Giant Airliner, then the DC3.

In those ships, between Pittsburgh and Newark, on a winter night, a man could learn a lot about flying the weather. It was about the toughest weather flying ever done. Instrument flying was new then. Radio aids and instruments were still quite crude, airplanes comparable in performance to present-day private airplanes. But the airlines flew almost the same weather as now. TWA had a milk run from Kansas City to Newark that took all night and made nine stops and some nights all nine required an instrument approach.

TWA made Bob Buck a Captain in two years. Then the war came and the first great surge of ocean flying, in four-engined land planes, under the aegis of Air Transport Command. Buck became Assistant Director of Training for TWA's Intercontinental Division (which was part and parcel of ATC). He checked people out on the DC4s and on the ocean routes. Next, for four years, he was the captain and manager of a special research project that TWA had taken on for the Army Air Forces. He had a B17 bomber ("Flying Fortress") to himself and a mission to seek out precisely the kind of weather that others stayed away from—the kind that gave the most trouble. It started with research into snow static—the kind you find best in Alaska. Icing research was added, then other things—before long, he was carrying 14 different projects. To find enough weather that was bad enough, Buck ranged from Alaska to Panama, out to Hawaii, to Southeast Asia, and, once, clear around the world. He ended up doing outright thunderstorm research. For this, they gave him a P61 twin-engined fighter (the "Black Widow") with lots of radar and lots of structural strength. He flew thunderstorms forward and backward, slow and fast, high and low. What was inside the monsters? Could you get through without spilling your gyros? How much did you dare slow up the airplane? What was the best way to keep control of the airplane? How bad was light-ning, hail, turbulence? It was one of the first deliberate, systematic series of flights through thunderstorms ever under-taken.

Buck then went back on the line, but kept doing new flight research for the military and the airlines both—on airborne radar and on *low,* low instrument approaches. When the In-strument Landing System (ILS) first came in, he was one of those who had to figure out the best way to use it. Once he wanted to study runway visibility in fog as it appears to a pilot breaking out of a low overcast. He let himself be hoisted up into the overcast, hanging in a parachute harness from a cap-tive balloon. He likes to fly.

He likes that left-hand front seat of an airplane. At one time they made him Chief Pilot of TWA. He found that the desk work interfered with flying, and he quit and went back on the line as a Captain. "The sky is my office," he once wrote. He was again tapped for the executive side of the company. TWA's president at the time talked to him about it and held out the prospect of a vice-presidency. Buck said: "Mr. Burgess, there are only two jobs on this airline I would want—yours or mine."

He owns a four-seat airplane with complete electronics which he flies to near places between trips to far places. He also owns a high-performance sailplane, which he flies in cross-country competition. He once made a private tour of Africa in a DC3. He has been around the world once sidewise —i.e., via both poles. He sits on various national and international committees that deal with piloting, is active in the Air Line Pilots Association, and acts as a consultant to manufacturers of airplanes and electronics. But always, when you see him again, he is just back from Paris or Bombay, with a cabinful of passengers riding behind him.

So that's the man who now comes in, as we stand there studying the weather map and debating what to do. He knows our problem. We are not engaged in an academic exercise: We are making, or passing up, a serious commitment. As we search the weather map we are also searching our soul: Am I good enough for this? Can I hack it? On paper, this weather situation can be dealt with by such-and-such a procedure. In reality, when the pressure comes on, will I get flustered and panicky, so that I can't do my best? Airplanes are flown in weather by real people, and the pilot—we, ourselves—is part of the situation.

Buck himself has been through enough tense situations to— well, to write a book. He's seen fellows' knuckles turn white on the controls—and maybe sometimes his own. There are some interesting short passages in his book where he touches on that side—panic control, self-control, the *constructive* use of the imagination. You can keep an easy touch on the control, he says, even though your knees are shaking. These passages will convey a little of the man who is talking to us—his air of ease, cultivated by long self-training and mental discipline. They will also give you the impression that he is your friend. It's interesting to note here some things he does *not* say.

He does not bore us, for example, with the phony good advice, "Don't Exceed Your Limitations. . . ." We do not know exactly what our limitations are. You don't know the breaking strength of a material until you have broken it. Nor do we know how tough this particular weather situation will be to fly. That's just our problem!

And he does not come at us with an advanced course in meteorology.

Certainly it helps, in flying the weather, to know some textbook meteorology, to have clear concepts of the things we meet in the air: cold fronts and warm fronts, cumulus clouds and thunderstorms, inversions, dewpoint and temperature, and types of fog, the principal air masses in our part of the world. This knowledge helps us understand the language of the weatherman and the meaning of the weather map. And it helps us to recognize these weather phenomena when we meet them. The nature of the brain is such that we see what we have seen before, and what we have a name for. We are blind to things which have not been properly introduced. People had fronts passing over them for thousands of years, but nobody ever saw a front as *front*—i.e., as boundary between contrasting air masses. Then, 50 years ago, the Norwegians first recognized the *cold front,* described it, named it. Now everybody can plainly see many frontal passages every year. In this sense, some descriptive meteorology helps the pilot fly the weather. But with the pilot's main problem— What do I *do*?—textbook meteorology helps only little. And just because it helps so little we are tempted to give him more —more than he has any use for: to go deep into the question of how weather is made. The Coriolis Force, the Geostrophic Wind, the Latent Heat of Condensation, the Adiabatic Lapse Rate, Frontogenesis and frontolysis. All that is a fascinating look into God's kitchen, but the pilot does not want to make the weather—or even the forecast. All he wants to do is fly it. And for that he does not need *more* meterology, he needs a different kind and he is getting it here.

What is it Captain Buck does for us? It's like untying a knot. The bafflement we feel as we try to judge a weather situation is a sort of knot in which everything is balled together. Weather. He shows us where to find the place to start and how to unravel it.

Read what he says about the Big Picture, the If-Thinking, and Way Out, and right away the problem takes on order. These are indeed things on which we can ask questions, find the information, and make judgment. Perspectives open, a strategy suggests itself. More questions follow. What will be the influence of the local terrain, the time of day? You feed that in, and you can make more judgments, now you have more to judge by: This mountain range okay now, probably will have many thunderstorms in the afternoon. Better go now. This city with this wind direction will have industrial smoke in the morning, but visibility will improve by noon, you gain by delay. And so on. By showing us what the productive questions are, Captain Buck arms us with a judgment capability we never knew we had.

Another problem that baffles us: how to monitor the weather once we are en route.

The weather here and now is okay. But where I'm going it is different, and by the time I get there it will be different again. Is the situation solid? Should I do something or just keep flying? How can I know?

Sure, we can get weather information by radio. But the same old problem: What is the intelligent question to ask? And what do we *do* with the information? Should we do anything? Many of us find it difficult to get an effective thought process started. We just keep on flying and hope the weather will hold. Buck reminds us that a whole weather system may not move as fast as "they" expect, or may move faster or may even back up. This is perhaps the most frequent reason for forecasts going sour. He thereby gives us another productive question to ask: Is the weather system moving as expected?

Now that we see the question, we can often get the answer we need by watching (for example, the progress of a cold front) at points which may be quite far off to the side of our route. It's really a quite simple idea. It's just a sample the captain gives us of what goes on in his mind when he is captain. But it greatly smartens us up.

Bob Buck writes as he talks and flies, with an easy touch. He uses small words. Unlike most professionals, he does not try to make his art seem mysterious and difficult: He makes it seem simple and plain. This might fool some reader into thinking that he is getting just a light chat. Not so. What we get here is in reality a sophisticated course in problem-solving. Buck shows us not *what* to think but *how* to think. Not "What should I do?" but "How do I go about deciding what to do?"

Rules never quite fit the real-life situation. But a man who acts on rational grounds and knows what his reasons are can deal with the variety of real-life situations realistically. And should things *not* work out as expected, should the judgments have been mistaken or the information false, such a man discovers his error early, while he still can do something about it. Rational decision-making: In business or government, we call this the Harvard Business School Approach. It is an American specialty, and it has been immensely productive. It will be productive for readers of this book.

Because of this sophisticated approach, Bob Buck's book is equally useful to pilots of all experience levels. All pilots do not fly the same degree of weather. Many people have to fly Visual Flight Rules (VFR), some fly instruments. Many have to be cautious, some can fly tough. But all have much the same problem of decision-making—the problem the captain shows us how to solve. And so, this book being about this sort of thing, and written by this sort of man, a pilot needs an introduction to it like a hungry man needs an introduction to a steak. Just start right in.

WOLFGANG LANGEWIESCHE

Princeton, New Jersey

WEATHER
FLYING

Weather Flying

WEATHER BOTHERS A PILOT only in a few basic ways. It prevents him from seeing; it bounces him around to the extent that it may be difficult to keep the airplane under control and in one piece; and by ice, wind, or large temperature variations it may reduce the airplane's performance to a serious degree.

That's what weather does. There are degrees and nuances, but all in all we fight weather in order to see, to keep under control, and to get the best and safest performance from an aircraft. The question is, How?

Well, we should know something about weather, what it is made of and how it moves. But a pilot wants to know the practical things, what to do about the weather. These practical things are a psychology for thinking about weather and methods for flying.

Prior to the developing of psychology and methods, a pilot should have one point firmly etched in his mind: the fact that weather forecasting is not an exact science. This statement is an old one, but it's true and ought to be thought about before tossing it aside as old hat.

The point is that the best weatherman in the world cannot forecast with perfect accuracy. The Weather Bureau can make impressive statements about how well they do, and their numbers are valid and true. In the overall picture they do a good job, percentage-wise; but the time they miss and pull their accuracy down from 100 percent to something less may happen to be the night our destination was forecast clear and fell on its face to zero-zero! Right then the pilot isn't much interested in statistics—except in trying not to become one.

This shouldn't be taken as criticism of weathermen. Actually a compassionate, understanding and friendly feeling toward them will do both pilot and weatherman the most good. What we are trying to say is that you cannot count on weather, because even with computers, satellites, and perhaps a little witchcraft, man simply cannot outguess it 100 percent of the time. In the simplest form we must realize that weather results from air-mass movements and that these are capricious.

The pilot's weather psychology has two parts. The first is skepticism. Being a weather skeptic is an important ingredient of the formula for living to a ripe old age. The second part is always to have an alternate plan for action. These two keys, skepticism and alternate action, are the foundation of it all.

Being a skeptic keeps us safe; having an alternate plan of action adds to safety, but more important it makes it possible to fly and to make the airplane work.

If we are completely skeptical, we put the airplane in the hangar and forget about it, and sometimes this is a good idea. But we are trying to use airplanes; we want to go places as much as possible. The alternate plan of action helps us to do so.

Say a pilot is headed for a place that's forecast clear; his skepticism says it's near the coast, night is approaching, and the temperature is near the dewpoint and the wind from the sea—it could fog in. Now the easiest way out would be to stay home: but he could also take off and probably make it, as long as he has enough fuel to get away from the coast, should it fog in, and fly back inland. That's alternate action.

That is a very simple example, but it is basic and important. It is the way the airlines operate and the way they keep going. They are skeptical about weather—skeptical enough to have an alternate plan of action in everything they do.

To fly in weather a pilot needs a certain ability and various degrees of equipment. Since every pilot cannot have all the ability and all the equipment, he must take on weather in amounts that fit his situation. This can be done, and the fact that a pilot doesn't have instruments, radio, and a rating to use them doesn't mean he has to stay on the ground whenever there is a cloud in the sky. But he must realize his limitations.

There is a point of confusion in this area that gets people in trouble: the mistaken idea that equipment makes up for lack of ability. A pilot can have an airplane with all the trinkets, bright and new, up to par and working well, but if he doesn't know how to use them, if he doesn't know how to recognize their limitations, he's worse off than a pilot with nothing more than engine instruments and a compass.

What all this means is that pilots must know and fly by limitations. All pilots, even 40,000-hour airline pilots, have them.

What we are going to do in this book is talk about flying the weather.

We will talk about weather in the meteorological sense and then about how we approach the weather problem and what to do about it. We want to talk about the pilot's emotions and about his thoughts when he finds himself, say, in the middle of a thunderstorm.

There are things to talk about such as instrument-flying techniques and how to test yourself to get an idea of your limitations. There are matters of technique, too, for the person who doesn't fly instruments.

One of aviation's greatest fascinations is the weather. When it's bad it consumes our flying thoughts, but we think about it, too, on a sunny clear day with a light wind and pleasant temperatures, if for no other reason than that we must say, "What a beautiful day to fly. I wonder if it will stay this way."

Weather has so many facets that we never stop learning about it, and more than 40 years have taught me for certain that I can never definitely say that the situation is pat and I know exactly what will happen.

We learn to respect weather and never to be complacent about it. This is the warm kind of respect we give to a beloved adversary, for, after all, weather gives us many things: green grass, a blue sky with fluffy white clouds, and the rush of a summer storm that thrills and excites us. It gives the cool soft kiss of gently falling snow, and the beauty, next day when the front has passed, of sparkling sunshine and crisp invigorating air.

Like a good friend, weather rarely bores us; it supplies constant variety for our lives. How dull it would be, on the ground or in the air, if we never had to ask, "What's the weather going to be?"

2 A Little Theory for Weather Flying

MOST BOOKS ON WEATHER start out by saying that air is made up of 21 percent oxygen, 78 percent nitrogen, and 1 percent other gases. This isn't that kind of book, and instead we might say that air is made up of wind, turbulence, clouds, precipitation, some fog, and a lot of nice clear days. But we cannot escape all the theory and must talk about some of it.

You will find lots of things said more than once in this book. This is because many weather factors apply to more than one weather condition and will automatically be repeated as different types of weather and flying are discussed. And lots of points are purposely repeated to be extra sure it's understood that they are important.

Weather is complicated. A deep study of the science shows many complex factors, but when you boil it all down the keys are temperature and moisture, visible (precipitation or cloud) or invisible (vapor). Basically there is always a certain amount of water vapor in the air, and when air is cooled, this water vapor is squeezed out and is made visible. Much of the study of meteorology revolves around the ways water vapor can bother us, and what processes there are to cool the air enough to make that water vapor visible.

THAT IMPORTANT DEWPOINT

There are a couple of items that pilots should know. One is dewpoint. Most of us know it as the temperature at which condensation begins. If the temperature is 50 degrees and the dewpoint is 45 degrees we have only to cool the air 5 degrees for the moisture to come out where we can see it . . . and if it's fog, that's all we can see. Dewpoint is handy for a pilot. It's on the weather sequence reports and it is a simple matter to look at the temperature/dewpoint pair and decide what the chances are for them getting together. That is a matter of deduction and common sense. If night is approaching the temperature is going down; if there's a body of water with a wind blowing from it toward the land, the dewpoint is going to go up, giving the same effect as lowering temperature. If it's early morning, it's obvious that the sun will come up and heat the air, separating the temperature and dewpoint.

A little warning, however: Sometimes fog doesn't form until the sun comes up. An airport may have the same dewpoint and temperature all during a still night and yet not fog in. Then, just as the sun is coming up and we think everything will be okay, the airport goes zero-zero in fog. The sun, one theory has it, creates enough turbulence and mixing to upset the delicate balance of the dewpoint/temperature relationship, triggering condensation. Of course, further heating as the sun gets higher will finally burn the fog off. If it's winter, especially in higher latitudes where the sun is low, this burning off may take time. So don't be complacent just because the sun is coming up. If there's a high overcast above the fog, it may never burn off.

If rain begins falling from a front it will raise the dewpoint, and if night is approaching it will lower the temperature; thus, both temperature and dewpoint are working to get together and make it a miserable evening. There are many combinations that bring the two together, and in most cases it doesn't take a scientist to figure them out. Dewpoint/temperature relationship is, therefore, a good guide for pilots.

HOW AIR COOLS

We should review, also, how air gets cooler. The simplest way is by the sun going down. As any ten-year-old knows, it gets cooler at night. This type of cooling is called "cooling by radiation." Another way to cool things off is to bring colder air to an area. A front goes by, the wind turns northwest, colder air flows from the north, and the temperature drops. Cool air also can flow to the land from a body of water. This process is called "advection," and it can bring in or take away moisture.

The other cooling process is called "adiabatic cooling." This is simply a physical law that says when air expands, it gets cooler. What makes it expand? Lifting. When wind pushes air up a mountainside, it is lifting the air. It goes up the mountain to a higher altitude where the atmospheric pressure is lower. The air cools by expanding at a rate of 5.6 degrees F for each 1000 feet it's lifted. If the air is lifted high enough, it may cool to the dewpoint, and then a cloud forms.

It's worth noting, too, that the opposite happens when air comes down a mountainside. The pressure increases, the air gets warmer, the dewpoint and temperature separate, and clouds or fog dissipate. The entire process can often be observed on a mountain. The windward side has a cloud near the mountaintop, beginning part way up the slope and hugging the mountainside and mountaintop with an eerie white cover. Then on the downward side we see the cloud shred off and disappear, leaving the downwind mountain slope clear.

We can see the same thing sometimes up high when watching a lenticular cloud downwind and away from a mountain where a wave has formed in the sky. As the air flows up the front side of the wave, which is the upwind side, a cloud forms: then as the air flows down the other side of the wave the cloud disappears. It's a fascinating cloud because it doesn't drift, but just sits in one place, forming and dissipating. One can watch the downwind edge of the cloud and see the motion as the pieces shred away and disappear. All this, the flow up or down a mountainside or wave, allows us to see the adiabatic process at work, lifting, reducing pressure, and cooling; lowering, increasing pressure, and heating; making cloud and then destroying it.

If the air being lifted up the mountain is unstable, the cloud does not dissipate; instead, it keeps going upward and may turn into the type of thunderstorm which we would call "orographic." That is the difference between stable and unstable air: Stable air comes back down when the force lifting it is removed; unstable air, once it has been lifted to the point where cloud forms, breaks loose from the lifting force and keeps going up by itself.

Other things raise air and cool it. A cold front pushes warm air up; it flows up over cold air and becomes cooler itself, becoming, of course, a warm front. And the reverse of nighttime cooling is daytime heating, which makes air rise as thermals, the things glider pilots look for to circle in and go up. When this air cools to its dewpoint we get cumulus clouds, and if the air has a certain moisture content and is unstable the clouds grow into thunderstorms.

A meteorologist sees all these changes and additions, or subtractions, in a sophisticated way. He studies upper air soundings and weather information from many places and in many forms and is trained to study a mass of information and analyze it quickly. He may even use fancy electronic computers to help.

We don't have all these devices, and if we did most of us wouldn't know what to do with them anyway. But we do have weather reports, and we can see. We can relate what we see to the big weather picture. We can ask ourselves simply, Is it going to cool off and is there moisture present, coming in, or possibly coming in, to go with the cooling? By doing this it is possible to reduce all the complex weather factors to a simple understanding. What makes a front potent? Warm air being cooled. What makes clouds on a mountain? Warm air being cooled. What makes fog over a seaside airport? Wind bringing in moisture, or cooling the air. What makes fog in the country? Moist air being cooled at night after the sun goes down. What makes low ceilings when it rains? Rain falling into lower air, raising the dewpoint. We can go on and on, and finally relate any weather that restricts our visibility to temperatures and moisture.

SEASON AND TIME OF DAY

In our thinking on temperature and moisture we should consider two important points: season and time of day. In the summer, things are more phlegmatic, and the weather is basically good, or it tries to be. In winter it is more violent and moves and changes quickly. But fall and spring are the most difficult times to predict. Air masses haven't decided whether it's winter or summer; temperatures can be colder or warmer than expected and give unexpected bad weather. The nights are not really long, but they are long enough to produce substantial cooling. A spring day can be mild and docile, or it can blow and be wild.

All our weather thinking should be related to the time of day. We must simply ask, is it the full part of the day when it is warm, or are we catching up with the cool night, when temperatures and dewpoints get together?

TERRAIN

Terrain is an important ingredient in the weather. Terrain which rises presents a chance for air to be lifted. Sometimes this rise in terrain—the orographic effect, as the big boys call it—can be very abrupt and dramatic. A mountain range may suddenly burst upward from flat ground, like the Rocky Mountains as one approaches Denver after flying over the miles and miles of flat land in eastern Colorado. On the other hand, rising terrain can be subtle, like the gradual slope of the land from the Texas coast of the Gulf of Mexico to the higher land in eastern Colorado. This rise sneaks up on us and doesn't clearly display itself, but the silent flow of warm moist air up this gentle slope can produce widespread fog, or kick off thunderstorms.

Terrain makes bad weather worse. A cold front being pushed up a mountainside is nastier than a front crossing Indiana, where the terrain is flat. Air-mass thunderstorms are kicked off more quickly when wind flows up a mountainside. Fog can form sooner in valleys where cold air collects. But to make things more cheerful, mountains can help clear up weather on the downwind side, where downflow heats the air and dissipates clouds, or keeps ceilings up. This effect often takes the steam out of cold fronts, making them more docile on the downwind side of a mountain range.

Air can lose its moisture on the upwind side of mountains and be dry and clear on the other side. A vivid example of this appears in the Far West, where the Pacific side of the mountain ranges gets a respectable annual rainfall and supports plentiful vegetation, while the eastern side of these same mountains is desert because most of the moisture is wrung out of the air on the western slopes.

To a pilot, this means that if he is over Los Angeles and the weather is bad he can fly over the mountains to the desert for good weather. All this, in its way, is the adiabatic process at work with terrain helping it.

And when we think of terrain we should not think only of mountains and valleys, but also of wide streams, lakes, and nearby oceans as well. Water and land generally have different temperatures. In winter the land is colder than the sea, in summer the reverse. You can see a demonstration of the temperature difference between land and sea when flying through the intertropical front along the east coast of South America. During the day the land gets hot, hotter than the sea, and great towering thunderstorms are everywhere—except over the sea, where it's cooler. So you fly out to sea in nice clear air. At night, however, the sea is warmer than the land and more and more showers are found offshore; so you fly over the land for the best ride.

The lee of the Great Lakes in winter has snow and stratus because the wind blows air across the Lakes where it picks up moisture. Then it blows up the slope of the Allegheny Mountains, rising. The result is zero-zero with snow and clouds on the maintains; the clouds are full of ice, and it takes 9000 to 14,000 feet to get on top.

Cities are a part of terrain–weather thinking. Cities make smoke and the microscopic particles from it are something on which fog forms. Smoke and air pollution make the formation of fog easier, and a wind carrying pollution toward an airport is a setup for poor visibility. That's terrain—manmade, but still terrain.

WIND

Another important factor in weather is wind, which plays a major role in a pilot's life. It affects us from the moment we take the airplane out of the hangar until we secure it for the night. Wind tells us how we must handle an airplane on the ground and during takeoff, it tells us how we must think and act while flying close to uneven terrain, it tells us how short we can take off and land, and what up- and downdrafts we can expect. It affects the performance of our airplane. A big jet weighing 290,000 pounds can take off from a certain runway in calm conditions; with a ten-knot headwind it can increase the gross to 300,000 pounds, but with a five-knot tailwind the gross is reduced to 280,000 pounds. The same rules apply for a Cessna 150; only the numbers are different.

But wind is also important in thinking about large-scale weather. First, on a weather map we notice, almost automatically, if the isobars are jammed together like tracks in a railroad yard, saying the wind will be strong, or if they are wide apart, indicating that the wind will be lazy.

Then we look at direction. East winds may bring bad weather, west winds sunshine. Knowing what the wind is, or catching its changes in velocity or direction, can give good weather clues.

Wind is layered and blows differently aloft than it does on the ground. The wind up high tells a pilot about speed for a trip, and so his range and fuel reserves. Wind just above the

The tremendous force of the wind drove this board (measuring 10 x 3 x 1) through a palm tree in Puerto Rico during a hurricane. (NOAA PHOTO)

ground, within the first 1000 feet, tells about shear and its hazards during takeoff and landing.

An important part of wind action is convergence or, more simply, places where winds from opposite directions bang into each other and pile up. The idea of convergence and what happens because of it is difficult to pinpoint, and the actions which it causes are complicated. A convergence area can be very big, like the intertropical front where northeast trade winds run into southeast trades and create an area of large cumulus and thunderstorms; or it can be tiny, where a sea breeze meets inland air and forms a miniature front of no special consequence except for a line of clouds a little way in from a coastline. These are called sea breeze fronts and generally are mild, but just to keep weather's ability to surprise us alive, thunderstorms occasionally will develop along such a front. Fronts are a demonstration of convergence, and so are low pressure areas. The important point is that almost any time convergence is present, there will be some sort of weather associated with it because of the process of air being lifted and cooled.

Divergence is the opposite of convergence. Air flows down and away and, going back to the adiabatic process again, heats up and generally gives good weather. A high pressure area is a large-scale divergence, a mass of sinking air.

This sinking air in a high, and rising in a low, affects flight more than we realize. Filing a flight plan in a heavy 747 from JFK to Paris, London, or Rome, I'd note whether I was going to be climbing through a high or low after takeoff before I'd decide what altitude to file with ATC for crossing Nantucket, a check point about 176 miles from JFK. If I were going to climb in a low I knew the airplane could reach 33,000 feet because the converging, rising air would help us climb faster.

But climbing through a high, with its diverging, settling air, would be slower, and so I'd file for only 29,000 feet at Nantucket, which was about all the airplane could comfortably make by the time we got there.

I'm sure most pilots, especially flying small, lower-powered aircraft cross-country in a fresh high-pressure area, have noticed how the airplane seemed to fly somewhat slower and work harder to keep normal cruising speed. This is worse in mountainous regions, but that's for another reason—waves—and we'll talk about them later.

A good pilot is wind-conscious; he trains himself to be aware of its direction and velocity, he knows how it smells and feels. He knows a warm humid wind or a crisp cold one, and where they came from and what kind of weather they will bring. He awakes in the morning, looks out the window, and sees on the ground where the wind is coming from, he looks up at the clouds, checks which way they are drifting, and learns the wind aloft. All through the day he is subconsciously aware of the wind and if it changes he senses it and asks himself what this may mean. He puts the wind and flying together, too, and visualizes it tumbling over some trees or buildings near the approach end of a runway and what that will do to his airplane. He tries to "see" the downdraft on a sharp mountainside; he always relates wind to aircraft performance as well as to weather. The good pilot is animal-like in his sensitivity to the wind, feeling and understanding its motions by instinct.

CLOUDS

A pilot literally looks at the weather to see what it's up to. One of the main things he watches is clouds. They tell a big story.

There are two cloud types, cumulus and stratus, and all cloud designations are some combination of them. There are three classifications: cirrus, nimbus, and alto. *Cirrus* are high-altitude clouds, and because they occur in high, cold air they are made of ice crystals; but they still follow the cumulus and stratus designations. *Nimbus* is simply a name given to clouds when precipitation starts to come from them—like cumulonimbus and nimbostratus. *Alto* simply designates height; it means a cloud is at medium height, somewhere between 7000 and 25,000 feet, and again it is used with the basic cloud forms, as altostratus, altocumulus. You never hear "altocirrus," because cirrus by itself is high.

The important part about the two basic cloud types is their action and this, in turn, tells how they were made. Cumulus clouds are bouncy clouds. They were born of instability, born in air that once it starts up wants to keep on going—for that's all instability is. Stratus clouds are smooth and flat, or almost flat; their air is basically stable.

Heavy precipitation comes from unstable clouds; steady, light rain or drizzle, from stable ones. Said another way, ceilings and visibilities will be high enough to land during unstable conditions, except that briefly heavy rain may cause the visibility to be near zero, the runway flooded, and stopping difficult. Precipitation from stable clouds means low ceilings. Light precipitation can bring zero ceiling, or near it, and the condition can be widespread and of long duration.

So, fluffy white clouds are cumulus, and flat, layered ones are stratus. To make it more confusing, they can be in combinations, as stratocumulus, for instance, which is a layer of clouds, containing some instability. The precipitation from the clouds of slight instability can be light.

The stories clouds tell are varied. Cumulus clouds are gen-

Three decks of clouds: At the top are bands of cirrus, probably part of a fast-moving jetstream; the middle layer, close to us, is altostratus left over from a frontal passage of some hours before; down low are cumulus which have developed in the unstable air coming in behind the front. This is the sort of story clouds can tell us.

erally thought of as pretty, fluffy white things floating in a
blue sky. They mean good weather. But they are not all the
same. We know that any cumulus-decorated sky will have
choppy air underneath the clouds and smooth air on top. If
we look at the clouds more closely we can get an idea of how
choppy it will be underneath. If they have a shredded look,
like cotton that's been pulled apart, it's probably rough, you
are slapped around the sky, and it's a good bet that the surface winds are strong and gusty. When we fly gliders in these
conditions, the rising thermals are generally chopped up and
difficult to stay in.

If cumulus are bulbous and fat, however, the choppy air
will not be so choppy and the up- and downdrafts will be
better defined. You rise and descend more like a boat in
swells at sea. We also look at these fat cumulus with more
suspicion, because they are the kind that may grow large and
a few of them turn into thunderstorms.

We can tell without even looking at a weather map, merely
from the type of cumulus present, a lot about the synoptic
situation. The first type, the shredded kind, are in an air mass
that's close behind a low; a front has gone by recently and
fresh, vigorous air is flowing into the area. We're in for a few
days of good weather. The fat cumulus say that we are deeper
into a high, perhaps on the back side of it, and warmer unstable air is coming in. Somewhere to the west a cold front is
probably starting our way.

Stratus clouds tell a different yarn. We may be flying in a
mountainous area such as the New England states. There is a
high overcast made of altostratus, the visibility is good. Our
destination, in the mountains further south, is reporting 8000
feet and five miles visibility with light rain—plenty good
enough. We know there's a rain area, a warm front approach-

ing, but the forecasts do not make our destination really bad until long after our arrival. We fly on and notice rain on our windshield. The visibility drops some, but there's enough. We are happy even though it rains a little harder. But then, looking down in a valley, we see a wisp of stratus below, just a little thin glob of cloud floating along my itself. But it's a red flare signal! Things are happening; enough rain has fallen into the lower air to raise its dewpoint, and stratus is forming—and stratus is the cloud low ceilings are made of. It's forming faster than the forecasts indicate; the next thing we know our destination will have about a 300-foot ceiling or less. We review our fuel, check the alternate and destination weather, which is going down, and wish we could hurry and get there before it socks in. All this was told us by a little piece of stratus.

We are flying westward on a summer day, on top in clear air with excellent visibility; below it's hazy and difficult to see. Way west of us there's a cold front, which is forecast to arrive at our destination long after we do. But suddenly our eye catches a different shading in the high sky far ahead. We take off our sunglasses to see it better but we don't; we put them on and squint a little trying to pick it out. We fly on and look some more. Then we're certain. The western sky holds solid cirrus, white and innocent-looking, but it's a sign that says let's check that front, it may be moving faster than we thought.

These are a few examples of the many things clouds tell us; they are an entire weather story put in the sky for us to read. We can study for a long time and never know the whole story, but it is profitable and interesting to try.

3

Looking at the Big Weather Picture

WEATHER AND FLYING go like this: check what it is and what it is going to be, and then watch closely while en route to see what it's really doing. These are the steps:

1. Get the big picture.
2. Digest the forecasts.
3. Check what it's been doing.
4. Get current information in flight and watch through the windshield.
5. Mix it all together for flight management.
6. On the ground, after flight, look back and see what happened.

THE BIG POINT, THE BIG PICTURE

The first item, the big picture, is called the synoptic situation, or general view. From it we obtain a picture of the pressure systems and fronts. This is the foundation for weather judgment.

Our knowledge and mental picture of the synoptic situation is important from the time we first check the weather before flying until we put the parking brake on and shut her down at flight's end.

The weather map and its information are old when we look at them, and by the time we use the information it will be older still. A very important point is that weather is always in motion. It moves and changes, and the movement and change are the things that make it necessary for a pilot to keep his eye on the action all the time.

A weather map is like a snapshot. The shutter clicked at 00:00 hours Z time; that was the moment everything stopped, but at 00:01 the map is different, and the characters in the picture have started to go their own ways again.

Surface weather maps are important, but at first look we can be confused by all the numbers and meterorological hieroglyphics. Actually these are mostly for meteorologists, and what may be interesting for us in those numbers can be obtained from actual weather reports. But there are important things we want from a surface map: where the fronts, highs and lows are; how the isobars curve, because they picture the wind flow and tell us where it's coming from, and what it's like, such as being warm and moving toward cold ground—which can mean a fog problem. The isobars also tell us the wind speed, because the closer together they are the stronger the velocity. And we learn if we will have head, tail, or cross winds. So the weather map is a general picture of where things are and, most important, how the sweep of winds will make them move and change.

The illustration on page 22 shows some features of a weather map in relation to the isobars. Notice the flow to the east of that low. The air is coming from the Gulf region and will put warm, humid aid over the eastern part of the country—and some thunderstorms too. Study how the isobars on various parts of the map show where the air has come from and what it's like.

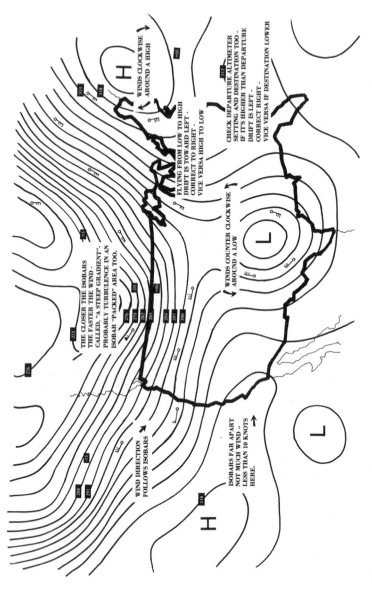

What isobars mean on a weather map.

But we are apt to think only of the surface map when, really, that isn't where we fly. Fortunately there are maps drawn for where we do fly, in the sky above. These are often overlooked and they should not be.

They show wind patterns, velocities, and temperatures at different flight levels. They aren't listed as altitude maps, but as pressure level charts expressed in millibars (mbs). Sounds complicated, but it isn't, and all we need think about is these millibar levels as altitudes. Various level charts are drawn, starting at 850 mbs, which is roughly 5000 feet. So if 5000 feet is where we fly, then a study of the 850 mb chart will give a more realistic picture of what the winds are and what's going on up there.

These pressure charts are made for:

mb	Feet
850	5000
700	10,000
500	18,000
300	30,000
250	35,000
200	39,000
150	45,000

There are higher charts for 100 mb and 50 mb, but we don't get them in most FSSs or weather offices, and we don't fly up there much anyway—that's Concorde country!

The altitudes are approximate because the maps actually show what level, in meters, we find 850 mbs, or whatever. So these charts, if viewed sideways, would undulate as the pressure levels do, reflecting lows and highs.

All we're doing in this little exercise is to show how the millibar level charts relate to altitude and to say that if we're

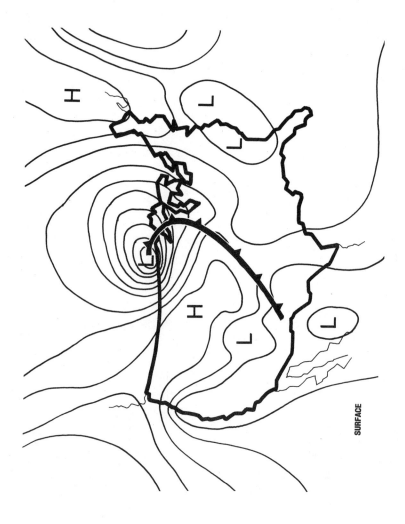

SURFACE

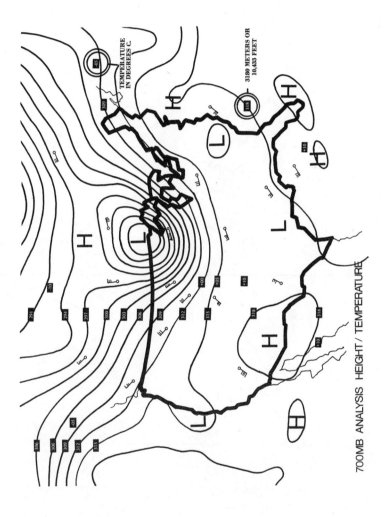

700MB ANALYSIS HEIGHT / TEMPERATURE

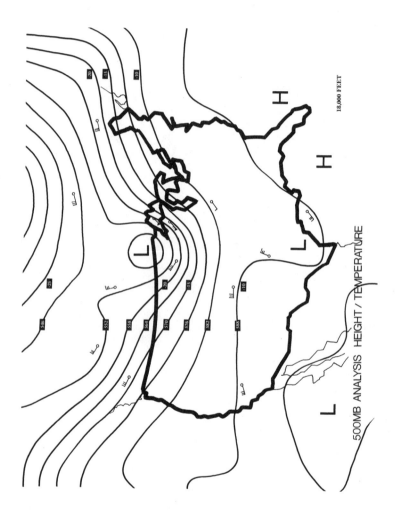

500MB ANALYSIS HEIGHT / TEMPERATURE

18,000 FEET

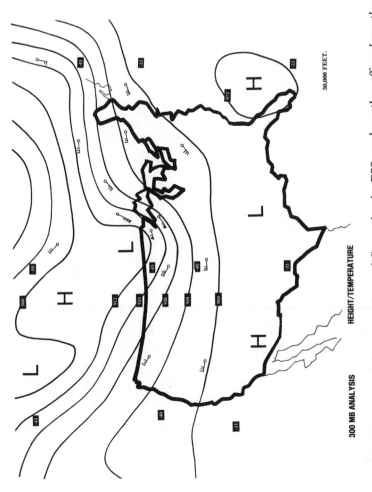

300 MB ANALYSIS

HEIGHT/TEMPERATURE

30,000 FEET.

Series of four mb charts—same day, same time—different levels. FSSs and weather offices have them.

flying near 10,000 feet we should study the 700 mb chart, and others for other altitudes.

The locations of highs and lows are different aloft than on the surface. At high altitudes one may not even see a low because it's all down underneath and the 200 mb chart, 39,000 feet, may have the isobars in a straight line with the low nowhere in sight. If it does show up at the 200 mb level, however, you can be sure it's a lulu!

The sequence of charts starting on page 24 shows the changes between the surface and 300 mbs. The surface chart has a deep low on the Canadian border west of Lake Superior. At 700 mb, 10,000 feet, we see the strong wind gradient on the south and southeast side. At 500 mb, 18,000 feet, the low isn't as intense and the isobars begin to become more west–easterly. At 300 mbs, 30,000 feet, the intense, tight circulation has smoothed out and started to join the zonal west–east flow. Most of the action is down low and the 30,000-foot plus jet airplanes will have little weather beyond ice crystal clouds and probably some light turbulence. These charts were drawn from actual charts for the same day and time.

According to the 700 mb chart, if we were flying from Seattle toward San Diego, we'd have a constant left drift because we would be flying toward a high, and require a plus drift correction, although not much because the winds are light along that route. We can judge the wind velocity by the isobars or the barbed arrows on the chart.

What we're trying to get across is that those upper-level charts are very important and not enough use is made of them. I always consider which chart to concentrate on in relation to the airplane I'm flying: If it's a small single-engine I look at the 850 and 700 mb charts, the charts for up to 10,000 feet. For a turboprop the 500 mb is the interesting one because

I'll fly in the vicinity of 18,000 feet, although some of these airplanes are getting up near 30,000 feet. In a pure jet I look at 300 mbs and up. Of course I take a glance at them all to see which way the lows are leaning, how deep they are and what's going on at other levels. Again, these charts are important because that's where we're flying and we should know what's going on up there.

But back to the big picture. It is often difficult to get to a weather office where there are maps. If there is a Weather Bureau or FAA Flight Service facility at the airport there are probably facsimile maps, or a map some enterprising Flight Service Station man has made himself, from teletype information.

But even without a map we can listen to general forecasts by the government's Pilots Automatic Telephone Answering Service (PATWAS) or on Transcribed Weather Broadcasts (TWEB), using the aircraft radio. There are a lot of ways to dig out the weather information if you want it badly enough.

NEWSPAPERS AND TV HELP

The newspaper and TV weather coverage gets better all the time and a newspaper weather map is often the best way for an isolated pilot to obtain the big picture.

An important part of checking newspaper weather maps is to look at the date and time the map was made. It may be pretty old by the time a pilot sees it, but looking it over carefully and then putting it together with weather reports from telephone or radio he can create a fairly reasonable weather picture. It's a good idea, also, to look back at yesterday's newspaper weather map and study the weather's previous positions and the rate of movement between the old and the new map.

LOOK BACK TOO

Even old maps have value. Once I stood in a weather office in Buenos Aires looking things over for a flight to Christchurch, New Zealand, by way of the South Pole. Since that isn't a regular route, the weather information was sketchy. The only weather map was a day and a half old. We made an educated guess, using the old map, that a front should go through Christchurch six hours before our arrival. We hit it almost on the nose, and after a 14-hour flight we arrived over Christchurch in sparkling clear air washed clean by a recent frontal passage. Not the way you like to do business, but much better than nothing. The big picture had helped.

TV weather charts are often more artistic than precise and you cannot be certain of the data's time. Some TV weather maps give the time and date of preparation as the newspapers do. These are the good ones. Some TV weathercasters are very conscientious and make up their shows at the last possible moment. TV reporting has improved greatly and in some cases outdoes the newspaper weather maps. But however we obtain it, we need to visualize the big picture and to get it firmly in mind.

We're going to stop here and talk a little about how one gathers weather and where it comes from.

There is a mass of weather data pouring out of Suitland, Maryland, our National Weather Center; fifty-one Weather Service Forecast Offices update the information and prepare terminal and area forecasts. There are all kinds of charts that crawl out of facsimile machines; terminal, enroute, and wind forecasts come via teletype as well as actual reports. It's certainly a system that produces and gathers enough information —and it's excellent stuff, as good as the state of the art can produce.

But the absurdity is that it's difficult for the pilot to get his hands on all this, make sense of it, and create a clear picture of what the weather is and will be over his route, and what he should watch for.

The weather people and FSS briefers try to be helpful, but because of the mass of data, pressures of ringing telephones and other demands on their time, such as working radio traffic, the pilot often gets only a generalized view of the weather. Because it is difficult for the pilot to dig out the weather story he can easily fall into the simple way by asking, "Is it VFR or IFR?" and, in return, receive a superficial response such as, "It'll be VFR," or whatever. Armed with that and some winds-aloft forecasts, our pilot is off and flying from a position of ignorance.

Let's face an important basic fact: A pilot must know the weather picture before he takes off and then keep up with it en route! I cannot stress enough the importance of a pilot understanding, for each flight, what the synoptic situation is, what kind of air mass he will fly through, and where the fronts are and how they are expected to move.

To do this he must dig out the big picture, the forecasts, and then *study* the actual weather reports along, behind, ahead, and to the side of his route for the past few hours so that, when he gets in the air, he can relate to each hourly report—which he should be gathering while flying.

Actual weather reports are tremendously important. They should be studied, as we've said, before flight for a number of hours back and then constantly obtained for every hour en route. Because weather forecasting is not exact, and probably never will be, we need the actual reports to combine with forecasts to know how the forecasts are doing and if they are giving us the weather promised or something different.

The actual weather report is the point of truth—all the charts, maps, and teletypes we've seen, or forecasts we've been given, are not *real;* they are only estimates of what will happen. But the actual weather reports are real! One can figure, guess, and get the opinion of experts on who's going to win the ball game, but we don't know who wins until the final score is posted. Well, the actual weather report is the final score! And what you see through the airplane's windshield is a final score too.

These things—the synoptic picture, forecasts, and actual reports—are the key. If this book only got that across it would be worthwhile.

Many years ago the teletype sequences went west to east and south to north, in sequence. That was logical because weather moves in these directions. You could watch a front move toward the east as you read each station weather report in line. The wind would be northwest at Kansas City, but still southwest at Columbia, Missouri, and you knew where the cold front was—somewhere between the two stations. Or one could see a warm front progress from south to north, graphically, because the reporting stations followed the same path. But today this logical concept doesn't exist; the teletype print-outs jump from one area to another, generally coming on a state-by-state basis, skipping around illogically inside each state. This system is set up for the convenience of the teletype circuits, not the pilot!

So what has this done to the pilot? It's made his weather study and en route updating difficult. It means he's apt to look up the destination terminal weather, but skip most of the en route stations that will help him visualize the weather's movement.

Since that's the way it is, the pilot must display great discipline and dig out what's needed himself until, if ever, the powers that be get this situation back the way it should be. All this requires the pilot to get to the weather station earlier and take more time. Also, a telephone briefing may be unsatisfactory because the man on the other end of the phone is showing signs of impatience when you ask for more information, especially when he just told you it would be VFR.

I was talking to a briefer on the telephone about weather from Philadelphia, Pennsylvania, to Burlington, Vermont. He said it was all VFR. But from the TV map I knew there was a front to the west that could move on course if it speeded up. I asked about it. "Oh, that won't bother you, your route's forecast VFR for your time period."

Not satisfied, I started asking for actual weather reports at stations like Buffalo, Rochester, and Harrisburg in an attempt to learn where the front really was. The briefer became angry and said, rather strongly, "What do you want those stations for? You're not goin' that way—I just told you it's VFR!"

Which demonstrates why people get in trouble with a sketchy picture of the weather. That front was moving faster than forecast and did mess up the course, and I had to go IFR after I started. But what about the VFR pilot new to the game and starting out on that briefing? Would he be smart enough to land? Turn around? Run east? Or would he press on, getting lower and lower, with less and less visibility in mountainous country, thinking it ought to be okay because the briefer assured him it would be VFR? It's the typical setup for a weather accident by pushing VFR too far.

This makes the point, also, that a pilot cannot depend on a

briefing being so accurate that a VFR-type statement covers everything. Weather science isn't accurate enough to do that, but the tendency of the system is to make it appear as though it is, and this means another adjustment in a pilot's thinking; he must not allow the official-looking maps, charts, forecasts, and briefings to lull him into feeling that with all that impressive stuff the statement, "It'll be VFR," is solidly accurate. In the air, in the weather, in clouds, ice, thunderstorms, fog or whatever, the authoritative appearance of the weather briefing office quickly fades away as the pilot combats the weather and tries to decide what this is that they didn't tell him about, and what to do about it!

Pre-flight study and constant updating of the weather will be his defense and aid when he realizes that the forecast and the statement "It'll be VFR" were inaccurate and the real world is different and tougher!

The difficulty of weather-gathering is rather frustrating because the briefer isn't at fault, he's busy, and his available information makes it arduous to present weather so that a pilot can be prepared to analyze it and its changes en route. Actual reports, once in the air, are incomplete and not obtained easily enough to get a broad picture of what's going on. This may change, from time to time, for the worse or for the better. The Enroute Flight Advisory Service (EFAS), and called on the radio as (for example) "Montpelier Flight Watch," is devoted to weather exclusively and is a quantum jump in weather aid to the pilot; he should use it freely. Unfortunately, at this time, it's only in service from 6 A.M. until 10 P.M. But regardless of how easily weather information comes to the pilot, it does not relieve him of the job of weather analysis and keeping up-to-date himself.

Knowing weather, and keeping up with its changes, a pilot

should realize that his flight route or destination may have to change with the weather. It is folly to plug along without the flexibility of thought, and willingness to realize that one may have to fly differently, change route, go IFR, divert, or even turn back!

These problems of weather information gathering and study, I believe, are the genesis of more weather accidents than any other single factor. The system has seriously reduced the pilot's ability and incentive to make weather judgments and be a flying weather man when, as we've said and will say again and again, a safe airman must be a weatherman too!

WHAT TO LOOK FOR

The best way to begin learning about the weather for the day we want to fly is to look at the weather map. At first glance it reminds me of a work of art; the sweep of the isobars and fronts is graceful and pleasing to the eye, as is so much of nature's work. But there is more than beauty; there's also an important story that tells where the air is coming from and what it's like—cold, hot, wet, dry—which, in turn, tells us about the weather.

When we see a deep low off the Maine coast, or one in the vicinity of British Columbia, with isobars on the westerly and northerly sides that sweep generally in a north, northwest–south, southeasterly direction, packed closely together, we know immediately that cold wet air is being swept into the area. The farther north the isobar north–south orientation goes, the stronger the air's cold-wet characteristic. There will be lots of weather in the area: low clouds, snow or rain, and wind—lots of wind. In the winter in New England the weather will be blizzard-like, and in the west the mountains will catch big snows.

On another day, another map, a northwest flow is over the midwest, with the isobars coming from continental Canada, and we know the weather will be excellent with fair-weather cumulus and good visibility.

So in that first look at the weather map we should notice the isobar trajectory and learn from where it brings air and what that area is like—cold, hot, dry, or wet—and how large an area it has crossed that will modify it.

Isobars connect points of equal barometric pressure, but they also show the wind direction, since the wind parallels the isobars to make a picture of the wind direction pattern. How close together they are tells the velocity; if they are tight—close—it'll be a windy day.

So it's the winds, along the isobars, that bring air masses to us. Arguments about what causes the wind and how all this movement gets started and pushed really don't matter to the pilot; he simply knows that the wind-isobar pattern is transporting air and it will mix, heat, cool, climb, or get pushed to manufacture his weather.

What do we look at when we see a weather map? There are two basic items, the pressure systems and the fronts. We look at them and visualize their movements and possible changes in relation to our flight path.

Unfortunately, sometimes we overlook the pressure systems to study the fronts. We forget, too, that a high pressure area, sometimes called a ridge, is also a system and not just a place between two lows. A high can often put out a lot of bad weather, and we ought to look at it with that in mind.

Where the lows are is also important. The fronts are a part of them and move or trail along as the low does. It is important to realize how far we will be from the center of the low. A long cold front coming out of a low in Canada and trailing back to the southwest has different weather in different places along its length. The weather at Burlington, Vermont, will be different from the weather at Harrisburg, Pennsylvania, and that will be different from the weather at Greensboro, North Carolina. Each front has its own character, too, and the weather is always different along them. There can be a lot of weather along a front, or none at all—such as a dry front. Sometimes the weather can be ahead of a front or behind it. All of which means you have to study the big picture plus each front's characteristics on sequence weather reports.

Weather is more intense near the low's center. In winter the ice is heaviest, the cloud masses thicker, more confusing, and more difficult to top or fly between; in summer the thunderstorms are wilder. If the course is near a low center we can be assured that things will be "interesting."

The frontal systems of a low are well known: a warm front, which generally moves south to north; a cold front which moves northwest to southeast; and, as the low gets older, an occluded front which rotates backward around the low. If the low moves, so do the fronts. If it keeps moving, its movement is easy to forecast, and a pilot will notice that things are working out as advertised. There are stationary fronts too, which are what the name says—fronts that don't move. They bring messy weather of fog and reduced visibility and, in summer, thunderstorms on a sporadic basis. These storms are difficult to see because of haze, fog, and cloud layers. Stationary fronts aren't violent, like fast-moving cold fronts, but they can cause problems because of low visibilities, especially at night and early and late in the day. The thunderstorms, once formed, can be tough.

Stationary fronts make a meteorologist's day trying because it's very difficult to say when the front will move, or whether it will just sit there.

WATCH THE SLOW LOWS

Lows that slow down are the nasties that really ruin a forecast. There isn't anything more difficult to figure out than a stalled front. If a cold front was supposed to go right through the East Coast but slows down and stops instead in the New York area, things go to pieces. Instead of clearing, the skies remain cloudy; wind hangs limply around the southerly quadrant, getting over to southeast perhaps; and fog and low ceilings prevail. With such a situation a kink may develop on the stalled front, a wave form, and a new low pressure area move up along the coast following the stalled front's line and putting out a lot of weather.

What this means is that if a forecast calls for frontal passage at 20:00 hours, but the front hasn't gone through by 21:00, it's time to be suspicious. Actually, to prevent surprise a good weather watcher will check a front's movement as it crosses the country, noting if it passes other stations on schedule. This way a slowing or accelerating tendency can be discovered in advance.

Again, wind is important. If it doesn't shift as expected, or its velocity changes, it's a warning sign that something different is happening. Winds picking up generally mean things are getting wilder. The rain or snow will be heavier and the air more turbulent.

If the winds slow down in our frontal system, we can worry about very low ceilings and visibilities—perhaps lower than one can land in at this stage of the weather-flying art—and they can extend over a wide area.

THE WIND SPEED TELLS A STORY

The difference between high winds and low and the resulting weather is experienced by airline pilots leaving the United States for northern Europe in the winter, a time when there is much fog. If winds are slack and there is little pressure movement they worry and take lots of fuel; but if a low pressure area is approaching the west coast of France, and Paris is forecasting 300 feet with rain and gusty winds, the pilot is quite happy and flies off without much concern; there will be enough ceiling and visibility to get in.

We become extra wary in slack winds and bad weather if we have one of the following: our airport near any body of water such as rivers, lakes, oceans, swamps, etc.; flight late in the day toward darkness; flight toward mountainous areas; flight to a place where there is a lot of moisture on the ground from snow cover in relatively high temperatures, which creates a surefire fog condition; ground soaked by previous rain; and flight toward cities and heavy industrial areas.

HIGHS ARE NOT ALWAYS NICE

As we visualize a low pressure area we often have a dark and foreboding feeling; thinking of a high brings sweetness and light and a feeling of well-being. As we said before, it ain't necessarily so. Highs contain, on occasion, fog, ice, thunderstorms, strong winds, turbulence, low clouds, and things we may not worry about directly but should, such as high temperatures that affect our performance or, in winter, very low temperatures that make an altimeter indicate higher than it should. A high has all these things and many of them depend upon where a low ends and a high begins.

Our low ceilings and poor visibility in snow over the Allegheny Mountains come with a northwest wind, which we think of as the front of a high, although you might call it the back of a low. The unstable air that a high brings to western Pennsylvania will build a cloud deck in winter that often extends as far west as St. Louis. On the mountaintops the ceiling and visibility will be zero; further west, where the air is older and modified, the tops will be lower and the bases higher. The instrument pilot will have a delightful trip on top despite battling some degree of ice getting up there and then getting down again; VFR it's tough.

Some of our wildest turbulence can be found in a high where strong westerly winds flow over a mountain range and cause standing waves.

Air-mass thunderstorms occur in highs. Most of them are scattered and easy to detour around, but a single thunderstorm becomes a major problem if it's sitting right over the destination airport. To make it confusing, air-mass thunderstorms occasionally line up and, for a while, give the appearance of a front.

The location where the back of a low and the front of a high meet is an important place. We tend to think that when the cold front has passed, the low is gone and we're now in a high, and that means fair weather. Well, it isn't always so, because the northwesterly flow on the front side of a vigorous high may be pumping in wet, cold, unstable air that's often a continuation of the low and its messy weather.

Highs, like lows, move, and when we see a high on a weather map that gives good weather, it's worth visualizing that it is in motion and will depart, and then what is behind it, such as a new low, will be moving in to take its place.

High pressure areas are often fog generators. The quiet, clear air near the center will cool, by radiation, to its dew point, and low land areas may well be fogged in from early evening until after sunrise. How early, how late, and how much depend on the season of the year and how long the high has lingered over the area.

Northern Europe is a great example of this: In the fall a high settles over Europe and often stays for many days (it's one reason that touring in Europe is best in late September and October—no rain). But the flat high brings nighttime fog, and as the days become shorter the fog burns off later and later in the day. In October fog forms in the predawn morning, and burns off by 9 A.M. or so. But by December it forms near midnight and may not burn off until noon. Some days it never burns off, and I once sat in Paris for five December days waiting for takeoff minimums! No particular hardship.

So while we're sitting in the middle of a tranquil high, airplanes are having a difficult time landing and may divert to various airports in Europe. Flying 747s to Paris, I've made many an approach to 400 meters visibility in this fall-winter high. A good aspect of it, however, is that in such flat highs the winds are calm or light; otherwise there wouldn't be fog. An instrument approach is easy to fly because there isn't any drift or shear.

The back side of a high—or we could call it the front of a low—has southerly wind flow that kills off the probability of fog because the air is warmer, doesn't cool off as much at night, and is moving. This inflow of warmer, more moist air is where the afternoon air-mass thunderstorms develop. And this air, moving from the south, runs up over the back side of the colder dome of high pressure and begins the process of the warm front of the next low pressure area. Milky, high cirrus clouds that dull sunshine, tell us, in an upward glance, that this process has begun.

So high pressure areas need serious attention too; what's

the wind, how old is the high, are we flying in its front, center, or back? It's all important.

DON'T FEAR WEATHER

All this may sound like the voice of doom speaking. It isn't meant that way. Weather shouldn't be feared, but rather respected. Weather makes flying interesting even though it frustrates us at times. But if we fear it our flying will be affected; we will not perform as well as we can and emotion may overcome judgment. It is fun, more often than not, to fly in bad weather. It makes us feel a part of nature again and not a coddled creature living in soft comforts; it builds our egos. The point is that to handle weather we must know, again and again, that it is capricious and its movements are unreliable. What we are trying to do is make certain that we outguess this uncertainty and, when we cannot do that, that we are prepared to handle it. The object is to have the last laugh.

—OR WORRY ABOUT IT

Worrying about weather shouldn't upset our lives. It doesn't help to worry in advance about what it's going to be like when you fly. I discovered, way back in my youth, that I'd start worrying about weather a day or two ahead. I'd study weather maps and try to guess what was going to happen on my flight day. I'd convince myself that there was going to be ice, or summer thunderstorms, and get worked up about it. Using a little extra imagination, it was possible to get quite upset about bad weather that was all in my imagination.

Then, somewhere along the way, perhaps as experience strengthened skill, I suddenly realized that there wasn't any way I could fly in weather until I climbed in the airplane and took off! From that time on I've lived with a philosophy of not being concerned about weather until I walked in the weather office or picked up a telephone to find out what it was that morning, that day, or that night.

Of course I watch weather every day to keep in tune with the big picture and its changes and movements. Simply watching TV weather will do it. This keeps me in sequence and gives a feel for how things are working before I fly. But worry? Never! I go to bed and to sleep.

WHAT'S THE WEATHERMAN SAY?

Just getting a picture of the synoptic situation isn't enough. We need to know what the weather people are saying about it —their description of it. This we get either from reading their analysis or from listening to it on a weather broadcast of some sort. Reading is best because it's more detailed. The broadcasts are shortened and edited, and often lose key points in the process.

The reason we want to get this analysis is to know what the forecaster is thinking, what he thinks the picture means and what he thinks it will do. Getting this information not only gives us an idea of what's going to happen, but what factors went into making up the forecasts. This makes our weather watching more useful.

"WHY?"

If we were able to talk to the meteorologist who made up the forecast we could ask him, "Why?" If he said the ceiling was going to lower, we could ask "Why?" and then he might say it was because of a front approaching. And we could say, "Why is it approaching?" and he might tell us about a low pressure area on the move. And we might ask why it's coming this way and he might explain that a high is blocking its movement toward a different direction and so it must move this way. And we could ask why the high is blocking and sitting still, and he'd have an answer for that. If we can ask enough "Whys" we can finally learn how solid the facts are on which he's basing his forecast, and thus decide how much confidence we want to place in the forecast. Asking "Why?" is a wonderful way to do this, and a great way to learn more about meteorology too. Unfortunately, we usually don't have a meteorologist to talk with; but we can ask ourselves the question "Why?" and try to figure out the answer as a good way of evaluating what may or may not occur.

FORECASTS

After getting the big picture we study the actual forecasts for the route to be flown to the destination and to an alternate or two, whether needed or not.

As we absorb the forecasts the big picture should form a background in our mind into which to fit the forecasts, and while doing this we develop our "if" thoughts.

"If" thinking is very important. We say to ourselves, What if the front slows down, What if it speeds up, What if the wind stays east, What will it be like if the forecast does a 180-degree turn? Being if-conscious keeps surprises out of life and has us prepared for something worse if it comes along.

A simple example is a cold front in eastern Ohio. We might be flying from Bridgeport, Connecticut, to Harrisburg, Pennsylvania. The forecast says Harrisburg will have scattered clouds, visibility five miles in haze, and southwest wind. It will stay that way until 17:00 when the scattered will become broken with a risk of thunderstorms. The forecast for later shows thunderstorms and a frontal passage.

It's no problem for us because we are going to arrive there by 14:00, long before the thunderstorms. But our "if-mind" says, Keep an eye on it. The front is out there and moving, a prefrontal line squall could pop out ahead of the front. It will be well to get the latest weather before leaving Bridgeport and current weather as we fly en route.

All this example does is demonstrate the kind of thinking a weatherwise pilot does, especially in relation to that synoptic picture in the back of his mind.

In this simple case it's obvious the forecaster based his terminal estimate on the movement of the cold front. It looks as though it will move, but will it move at the rate expected? Best information says it will, but as we know, they cannot hit frontal movement on the nose. The forecaster planned for that a little in the report when he warned of a risk of thunderstorms for 17:00 in case things moved faster than expected.

If we were coming from Dayton, Ohio, to Pittsburgh in this case we might be expecting Pittsburgh to be in good weather behind the front, but if it slowed down, Pittsburgh would not clear out and the front could still be in the area for our arrival.

It's all part of the necessary suspicion, the constant knowledge that weather forecasts do go sour.

So when we look at the weather we keep in mind several key points: big picture; time of day and season; temperature-change possibilities; moisture-change possibilities; terrain; wind; forecasts; and the big WHAT IF!

Checking Weather Details

NOW WE'RE READY to look at the details of the weather that will affect our flight, but as in the big picture these details are not always easy to obtain.

The degrees of availability of weather information range from the case of a pilot present in a national weather service office with all facilities, including a meteorologist, on hand, to that of a lone pilot far back in the bush tightly clasping a headset to his ears as he tries to hear a broadcast from some distant FAA station. In between these extremes is the pilot who can either visit a Flight Service Station, call one on the telephone, listen to a recorded FAA telephone message, or possibly, on a fee basis, use a commercial weather service, plus TV and newspaper coverage.

It is practically impossible to describe how to get what's needed in each of these various situations; but if we talk about studying weather as though we were in a weather office with a meteorologist, we can see what we'd like to have and then use our ingenuity to create a weather picture with whatever information is available. Often this involves some detective work as we use clues and deduction to fill in where information is lacking. It can be frustrating work, but it can be interesting too.

HOW TO LOOK IT OVER

There isn't any special sequence to follow in looking at the weather in detail, but I suspect that most people begin with the destination; what is it? Is it VFR or instrument and how close to each? It doesn't take too much soul-searching for a pilot to judge if weather is within his ability and equipment. Basically the flight either can be made VFR or it requires IFR.

Often, if one can only go VFR, he has a more difficult decision than the pilot who's able to go IFR, and this is an important point. The VFR pilot must stay that way, and this means judging if it will be VFR all the way, and stay that way. He has to scrutinize the forecasts, think about en route terrain, study actual reports, and be aware of the time of day. Sometimes it's a demanding decision that's being made by the least experienced pilot! But he should make it himself and not be satisfied by a briefer's simple statement, "It's VFR all the way."

The IFR pilot doesn't have to worry about clouds en route or terrain, because he can fly at safe altitudes, on instruments, above it all. He doesn't fear flying in clouds—IFR is within his ability. But the VFR pilot who doesn't know instrument flying has a very strong concern: he must not get into clouds or areas of reduced visibility that do not give enough reference to fly the airplane or see what's ahead, such as mountains, TV towers, and what not.

Once a pilot looks at the weather and says to himself that it's good enough, the question is whether or not it will stay that way. How does he answer that? Back to the big picture to see if the weather approaching is in the form of some kind of front; or if a front recently has passed, indicating improvement; or if the airport is near the center of a high, which affords protection on both sides.

The destination isn't the only area of interest, however. We want to consider alternate airports for both our destination and departure point. Either VFR or IFR we may want to return and should know if the departure field will be accessible, and if it isn't, where we can go in case of trouble when we are still close to our takeoff point.

Regarding airport weather, then, we are interested in destination and alternate, departure and alternate.

If the destination is too bad, then a pilot wants to know when it is going to improve. What do we look at first? The forecasts. The best way of studying them is to check them over a period of time. We check for the current time—that is, what the forecast predicts for this very moment—then for the previous four hours, and finally for the period of our expected arrival and a few hours after that.

TEST THE FORECAST

With all this firmly in mind we go to the actual sequences and see how the forecasts have been performing. What *was* the weather during the previous four-hour period, compared to what the forecast said it would be? Then the present: What is it actually doing compared to what the forecast says it should be doing? Now we have a feel for the weather, for the kind of weather it is and whether or not it's acting according to plan. If it hasn't been doing as expected, we'd best look again at the big picture and try to decide what's been going on. Are the fronts moving faster or slower, are the lows slowing, has a flow of air from the sea remained instead of turning around? The important point is to give the weather forecasts the test and to see how they have been performing. Good performance can indicate an easily forecast situation. A bad forecast generally means a tough setup that has the weatherman guessing. We learn to look at forecasts with a degree of confidence based on their past performance.

Once we have seen what the weather has done and what it is now doing, it's only necessary to fill in what it will be doing when we get there.

THE LATE WEATHER

There's another point worth consideration: what our airport is forecast to do after we get there. We should know the expected weather for a period of four hours after our arrival. This isn't in case we're late, although that could be important, but rather to give a picture of the trend. If, for example, the airport is forecast clear and unlimited for a long period after our arrival, we feel more relaxed about the chances of its being good when we arrive. If, however, it's forecast to begin a gradual deterioration four hours after our arrival, then we keep an eye on the possibility of this deterioration occurring much earlier than expected. What we are doing is bracketing the expected weather for our arrival; we get the forecast for *before, at,* and finally *after* our arrival. What we do is slip the time of arrival that we will really use in between the now-weather and the future-weather, and see toward which time—now or later—it actually develops.

REGULATIONS AREN'T THE IMPORTANT CRITERIA

The things we look at to start with are ceiling and visibility. Are they forecast to be within VFR limits or will an instrument flight be necessary? This question involves the Federal Aviation Regulations (FARs), which say that you need a certain minimum ceiling and visibility to fly. What these minimums are is determined by regulations; but an important point about legal minimums is that they are a batch of words in a book and aren't to be considered a substitute for good judgment. An airport may be forecast to be better than VFR limits and still not be a place to fly. Suppose it's down in a valley surrounded by mountains that are cloud-covered; there just wouldn't be any way of getting in there VFR. The legal minimums may not always be good enough for some situations.

When looking at ceiling reports, remember that the ceiling reported isn't for the scattered clouds, but for the broken and overcast. So if a ceiling is reported as 800 feet with scattered at 200, realize that you may have ground contact at 800 feet, but there will be annoying scud occasionally blocking your visibility much lower. If it's raining, you can bet that the scattered will become broken or overcast and become the 200-foot ceiling later on.

Visibility is an important factor for all pilots, whatever their experience. If ceilings will be at our near minimums and the visibility is good, then we know it won't be difficult to fly; but if visibilities are low, a minimum ceiling makes flying much tougher. If we are poking around mountainous terrain with low ceilings and low visibilities, we are in a difficult situation. A visibility of five miles with a 1000-foot ceiling doesn't sound bad in Indiana, but that same weather between Winslow and Kingman, Arizona, would be very hazardous and difficult.

The same applies to the low approach. A 200-foot ceiling with a couple of miles visibility isn't a difficult approach if turbulence and wind are reasonable; but when the visibility drops to half a mile, the approach is much different.

Conversely, we can handle reduced visibility if there is a lot of ceiling so that we can get up high enough to clear all the terrain we cannot see. In practice, however, we prefer visibility to ceiling if we cannot have both.

Whenever we consider visibility we should note whether or not there is precipitation. Four miles reported by the Weather Bureau can be much less when seen through a rain-smeared windshield. Four miles may in reality be only one bleary, wavy mile. Airplane windshields vary in their ability to shed precipitation, and some are really awful. Many aircraft are now being equipped with windshield wipers, which help tremendously. I don't think we realize how much they really do help. A pilot may not think he has any problems, because he is able to see through his windshield in rain; but if he could suddenly turn a wiper on and then look, he'd realize that he's been cheated out of a couple of miles of visibility all the time he has been without wipers. So the airplane windshield and how well you can see through it in rain is an added weather factor.

POLLUTION AND VISIBILITY

Visibility is a function, also, of smoke and haze coming from industrial areas. Perhaps an airport is reporting reduced visibility, while the general weather conditions are good. The reduced visibility may not be cause for alarm, because it could be industrial air pollution drifting in from the city. A look at weather reports from other airports that are not downwind of a city will show the real visibility trend.

With a light west wind, which says basically that the weather ought to be good, LaGuardia airport at New York could have one-mile visibility, when Newark airport, upwind of smoke-producing areas, may have seven miles or more.

An airport's location can mean a lot. If Long Island, New York, has a northeast wind and there's worry about fog and low ceilings, it's a cinch that LaGuardia Field, on the north shore, will have worse conditions than Kennedy on the south shore. The wind has to flow slightly down and over land to reach Kennedy, and this will help to raise the ceiling. What we're saying, again, is that we must include the factor of terrain in our weather evaluation.

If we are going to fly instruments into an airport, we can accept a forecast that calls for weather within legal limits for an instrument approach; but these limits, again, are only legal verbiage and that you are legal doesn't always mean you should go. Legal means regulations with fixed numbers like "200 feet," but numbers cannot cover all the factors in weather flying. Writing the regulations is a tough task. Wind and turbulence can affect an approach and may make a 300-foot ceiling seem awfully low. In an extreme example, you might have a legal 300-foot ceiling forecast for Daytona Beach, Florida, during a hurricane! You certainly wouldn't want to go there. To some degree, these elements enter into all weather judgments. If it's going to be necessary to descend through a thick layer of icing and a pilot's experience and anti-ice equipment are limited, he should not be interested in making an approach even if it is legal. There are a lot of common sense factors in judging whether to fly or not, and legality shouldn't always be the deciding one. Of course, a pilot cannot operate *below* legal minimums unless he's ready to declare an emergency or pay fines.

There are other times when the legal minimums look too restrictive—for example, a low stratus condition with calm air and little wind. In such a situation a fellow could probably cut a ceiling below legal minimums without much effort.

Generally, however, legal minimums are something not to go below, and there are times when one should remain above them, times when they are too low.

Cheating on minimums isn't clever. It's well worth remembering that minimums are based on many things, and the important ones are terrain and obstructions under the descent path, to the sides of it, and in the missed approach area. Sneaking below minimums means taking the risks of hitting something.

HOW DO YOU FEEL?

Another serious factor in weather judgment—often overlooked—is how a pilot feels. Some days we feel good and very tigerish; we've had good rest, our physical condition is tip-top, we feel able. At other times we may be tired or have a cold that lowers our efficiency, or we may have a hangover. We may be unhappy because of anything from a bad business situation to a fight with our wives. It's an honest fact that we aren't the same every day: athletes aren't, and they play, fight, or compete better on some days than others. It is difficult to stand off and objectively analyze how we feel; but some days you've got it and some you don't and the ones you don't should, under some conditions, raise your ceiling and visibility limitations.

I can remember an all-night flight from Kansas City to New York with stops at St. Louis, Indianapolis, Dayton, Columbus, Pittsburgh, and then Newark. It was back in DC3 days, and the weather was bad with an instrument approach at each station down through an icy overcast to minimums. The flight was a heavy mail flight, and at each station we were delayed while they loaded and unloaded mail. I had a new copilot, too. Finally, 5 A.M. found me in Pittsburgh looking at Newark's weather, which was forecast poor with precipitation. I turned to the agent and told him the flight would hold for six hours and quickly added that it wasn't that the weather was too bad, but only that I was pooped and didn't think it wise to fly any more bad weather without some rest. The airline, incidentally, never complained about it, and when I dropped by the Chief Pilot's office a few days later to tell him what I'd done, he complimented me on the decision.

MORE ABOUT WIND

We've looked at the forecast, the ceiling, visibility, precipitation, and at how we feel . . . what else? Wind direction and velocity in relation to terrain, such as light winds flowing from bodies of water or strong winds flowing down nearby mountains. Some airports, snuggled beautifully against the side of a mountain, can be a tough place to land with high surface winds.

There are details, too, of runway alignment. Now and then we may be headed for a field with a single runway where a crosswind will be too much to handle.

High wind velocity is an indication of turbulence in a low approach, which will make the approach more difficult.

Calm winds and approaches are delightful; they are easy approaches even though, quite probably, they will be low ones with reduced visibility because of the calm air.

ALTIMETER SETTING

Altimeter setting is reported on the weather sequences, and as we look at present and past reports, it's a good thought to make mental note of the pressure and see if it's increasing, decreasing, or staying the same. We can catch a weather trend this way and, keeping in mind the pressure, we have a reference for altimeter setting in case we don't hear one along the way. This is not recommended for IFR flying for which accurate altimeter setting is needed all the time; but the information is a little backup "just in case."

By noting the departure altimeter setting and the destination airport setting, we can get a rough idea of the drift along course. If the destination setting is higher than the one at our takeoff airport, we're flying toward higher pressure and will be drifting left and need a correction to the right, a + correction. You have to consider the possibility of a change caused by a frontal passage before you get there, but being observant of the settings will help and we'll talk more about it later on.

TEMPERATURE AND DEWPOINT AGAIN

On those weather sequences we find our old friends temperature and dewpoint. We want to keep them in mind to relate to the present and forecast weather, and as a reference when watching the trend at our destination. It's important to note, again, that time of day affects the value of temperature/dewpoint.

PILOT REPORTS

The pilot reports and remarks tacked on the tail end of sequence reports are worth careful consideration, because another pilot reporting the height of the tops, for example, is the next best thing to being there oneself. Like almost everything else, however, these reports must be absorbed with some thought about their validity. A report of the tops is a pretty definite thing, but reports of turbulence can be questionable. How turbulent is it, really? One pilot, being the nervous type, may yell severe turbulence while old hard-nosed Joe says it's just choppy. Notice, also, what kind of an airplane the report came from—one with a high wing loading wouldn't be as troubled as a lighter airplane of the two- or four-place variety. So, what do we do if turbulence is reported? Look over the general situation and decide if there should be turbulence; if there should and it's reported, that's that. If there shouldn't be turbulence and it's reported nevertheless, then either a Nervous Nelly was flying the reporting airplane, or something has started to move in the weather pattern that's different from what's forecast, and we'd better look things over again and decide what it might be. In either case the turbulence report didn't hurt, and has only caused us to study things a little more and be prepared.

As pilots, we should make an effort to report the weather we find, particularly cloud tops and bases, icing, turbulence, thunderstorms, or anything unusual. These reports not only help pilots directly, but they help the weatherman report and forecast, which finally helps pilots and a lot of other people too. Weathermen have information on places that report weather, but they do not know what's in between. Sometimes weather will cook up something they never knew about. A pilot has a chance to see these in-between places and can suspect that something different from the report might happen. This is the time—an important time—to make a pilot report and to let that weatherman know too. A report that says, "Snow encountered between Albuquerque and Gallup," may be an important clue that things are happening faster than expected, and the weatherman had better revise his forecast.

SUMMING UP

For the terminal, then, we study forecasts for past, present, and future; we read old and current sequences and check them against the forecasts to see how the forecasters have been performing. We look at other terminals within a couple of hundred miles of our destination and relate them to our destination weather, and most important, we study reports and forecasts of airports beyond our destination from which weather might come. If we are westbound and a front is approaching the field for which we're headed, we check how the front has been acting as it approaches. Say we're going from Columbus, Ohio, to St. Louis; it's important that we study, in addition to St. Louis, Kansas City, Springfield, Missouri, Oklahoma City, and Burlington, Iowa, and a station or two west of there, like Omaha. A wide sweeping glance will tell us where the action is, and then we relate it to St. Louis.

We need an alternate for our destination airport, even when flying VFR and especially marginal VFR. We study alternates as thoroughly as we do the destination, and perhaps add a few factors. One is that we want the alternate to be as good, and have as little chance of going bad, as possible. While we may head for a destination that has a nervous forecast, we want an alternate that has a solid one, and we don't like alternates that haven't.

The FAR can sucker one here, because their alternate minimums are fairly low and it is possible to have an alternate that is forecast to be above alternate limits for arrival and below them an hour later. This brings about such statements from pilots as, "That's the paper alternate, but my real out is ———." They want something that's definitely going to be good, like an airport that is behind a front and on the uptrend or one that will not be influenced by the destination weather, like Montreal when the New York area has a low moving off the coast. Long Beach, California, may be a legitimate alternate for Los Angeles, but in his mind a pilot says that Palmdale is where he can really go. It's wise to consider that the FARs give numbers to go by, but since weather isn't as precise as numbers, one needs extra protection on many occasions above and beyond the FARs. Paradoxically, or hypocritically, the FARs are often too restrictive and hold us back when it isn't necessary. However, this is the outcome of an interplay between the need for definite rules in the book, and the fact that so much of flying and weather depends on judgment. Judgment is difficult to cover in regulations. I'm glad I don't have to write them.

It's worth noting, before we leave terminal weather and the ways of looking it over, that we always relate our weather study back to the basics: big picture, time of day and season, temperature change, moisture change, terrain, wind, forecast and the almighty "IF."

5 Checking Weather for the Route

AFTER CONSIDERING the destination and deciding it's good enough, we have to learn what's in between: the en route weather.

What bothers us en route? If we're VFR it's cut and dried: we must have enough ceiling and visibility to stay VFR, and the more mountainous the terrain the more ceiling and visibility we need. We can fly on top of clouds, but they must be no more than scattered and very clearly and confidently forecast to stay that way!

On instruments our concerns are ice and thunderstorms—tops, bases, and ceiling underneath, especially if we're single-engine and want some degree of tranquility regarding the possibility of engine failure. Turbulence is a factor, but is generally related to thunderstorms. Wondering about ice and thunderstorms, the important question is: Are we dealing with these two things in fronts or only air-mass conditions? This knowledge is very important!

VFR or IFR, of course, we want to know about winds—head or tail and how strong. But more of all this as we go along.

The big picture has its usual importance, and we want to take note of the weather system that may move onto or off the route. The method of looking over en route weather is much the same as for the terminals; we want the same scrutiny of forecasts for before, now, when we get there and after. We mix in the same factors and emphasize terrain because it intensifies any bad weather that may lie along the flight path.

WEATHER IS MOSTLY GOOD

We talk a lot about bad weather and may put a gloomy look on everything, but we'd like to make two points. One is that if we only talked about clear weather there wouldn't be much to talk about. The other is that most of the time, thank heaven, the weather is good. We fly in clear skies more than we do in annoyingly cloudy ones. We may have cloud cover due to postfrontal conditions, but it is easy to top and fly above.

Fronts along the route, with their associated low-pressure areas, are the things that can make it tough. In these areas we find the difficult weather, the ice or thunderstorms, thick cloud decks, high-speed winds, and turbulence.

But this kind of weather covers our routes only a small percentage of the time. I spent four and a half years in weather research, trying to find bad weather. I sat on the ground many, many more hours waiting for bad weather than I have ever sat as a pilot waiting for good weather. Often, too, the bad weather that was supposed to be out there wasn't and instead of ice or snow I'd find myself flying disappointingly on top looking at blue sky or stars. All this is simply to say that things are good more than they are bad; but we have to talk most about the bad.

Clouds rarely start at 200 feet and go up in a solid mass to 30,000 feet or so. We can draw a mental vertical section of how clouds are stacked. Usually they are in layers, which become more complicated and the spaces between them fewer, the closer we get to a low or front. In the simplest form we may have a single stratus deck behind a cold front, one layer with a reasonably reachable top or one layer where warm air is overrunning cold ahead of a warm front—a cirrostratus deck, for example, with a high base and nothing below. Then, closer to the warm front, we find an altostratus layer with a lower base. Closer in this lowers still further and rain starts, whereupon a low stratus deck forms, adding another layer. At the front, the layers merge.

SOMETHING ON FRONTS

A cold front has much the same profile, but its area is smaller. The distance from front to back is shorter and it does not take long to fly through the front. A cold front can, however, be more violent than a warm front. Warm fronts are slow —and sometimes tormenting—while a cold front is a quick punch in the nose.

Fronts, as we know from primary meteorological study and from picture-book drawings, come out of a low pressure system. The warm front is a long arm sloping out ahead of the low, and the cold front is another arm pointed down, southward, from the low. The farther away from the low we are located on a front, the less violent the weather; and the closer to the low, the tougher.

OCCLUSIONS AND ZIPPERS

A type of front we haven't talked about is the occluded front. I remember, when I was a very youthful copilot who hung around the weather office on my days off, asking the meteorologist what an occlusion was. He said, simply, "It's just like closing a zipper." This really didn't tell me much until I learned more and found that his explanation was quite exact. At the center of a low the warm and cold fronts meet. When, near the center, they get together and one catches up with the other, they form one front. This process of catching up progresses downward from the low center. It's as though a zipper pullhandle were at the low center, with one side of the zipper the cold front and the other the warm front; the zipper closes southward as the cold and warm fronts come together. What does it all mean? Generally, it means the low is beginning to fill up and weaken, with cloud bases getting higher and tops lower, until the low is finally a trough with little weather. But in the early stages of an occlusion things can be rough and tough, with the characteristics of a cold or warm front depending on whether the cold air is catching up to the warm front and lifting it upward as a cold front, or riding up the warmer air ahead to give it the qualities of a warm front.

Sometimes the "closed zipper" portion of the occluded front tips and bends over backward. It is then called a "bent-back occlusion," and will act like an additional cold front. Because of the circulation about the low, such an occlusion feeds itself moisture and can cause some mighty mean weather.

Generally we think of an occlusion as "occluding out," which means that the low is starting to fill and die. Like everything in weather, however, you cannot always count on this, and while an occluding low generally spells gradual improvement, it can, on occasion, regenerate itself and create more bad weather. It's consequently necessary to keep a close eye on occlusions.

An occlusion is, in one sense, a trough. Troughs are areas between two highs and have lower pressure and some of the conditions found in a low. The conditions aren't as violent as in a fully developed low, with only high clouds or a medium deck, a wind shift, and perhaps a shower or two.

LARGE AREA WEATHER

Some frontal conditions can be very widespread and cause weather over a large area. We find this most often in winter, in big lows having large areas of overrunning warm air in their northeast sector. In winter, these can cause large areas of freezing rain. Usually this type of storm is not difficult to spot or predict, and the Weather Bureau will have them sufficiently well figured out for the problems to show up in the forecast. As far as flying into such an area is concerned, we shouldn't; it's better to consider the coziness of a fireplace and a good book.

THE IMPORTANT NE CORNER

As we study weather systems it's important to visualize where the weather is. In a low, most of it is ahead of the system, in the northeast corner (in the northern hemisphere). This is ahead of and in the warm front. If we were to cross a system starting in the east–northeast and to fly southwest, we'd first fly under high clouds and then into an area of rain or snow or possibly freezing rain, and then through an area of heavy precipitation, ice, and thunderstorms. This is the warm frontal surface. Then we would break out in a relatively clear area with a deck of cumulus or stratus or even clear skies. This is the warm sector. Incidentally, a low will generally move in the direction in which the isobars are oriented in the warm sector.

Past the warm sector we'd bump into the cold front: a nar-

row band of thunderstorms or ice, with lots of clouds and turbulence, but a fairly quick passage. Behind the front are
cumulus or stratocumulus and gradually, as we fly further,
clearing skies.

How fast all this occurs depends on our distance from the
low center.

The interesting part of the system, however, is that northeast
portion. In this area are the easterly surface winds that cause
fog and low clouds. Toward the front, precipitation begins,
causing low clouds and wide areas of poor visibility and ceiling. In winter, rain aloft will fall into cold air and create freezing rain and ice storms. This is also the area of heavy snowfall.
This area may often be extensive enough to make finding an
alternate airport difficult; it can also take a long time to move
off. Although we cannot forget the cold front, the big action
is up there ahead of the warm front, and this can be true
hundreds of miles away from the low as well as near it.

In Dr. Horace R. Byer's book, *Synoptic and Aeronautical
Meteorology,* he says, very accurately, "The warm front situation may properly be regarded as the most serious hazard to
aviation. . . ."

That's a good statement because a warm front has it all, and
it's not simply the frontal surface itself, but also the big area
northeast of it, that requires study and respect when we look
over the weather.

Warm fronts slope, in a shallow way, from 100 to 300 to 1.
So if the frontal surface—the place where surface reports show
wind shifting from easterly to southerly—is at, say, Amarillo,
Texas, a pilot will find the warm front aloft anywhere from
100 to 300 miles northeast of Amarillo. That would be the
altitude where warm front thunderstorms begin, or how high

we'd have to climb for above-freezing temperatures when there's freezing rain down low and we want out of it. This warm front slope is worth mulling over at leisure, and important to note when checking weather.

GO THE SHORT WAY

From studying weather, we tend to get the idea that we are always flying through fronts at right angles to them. Of course this isn't the case; sometimes our course is up or down a front, or at any angle to it, in which case tough weather can be prolonged. So there are times when it's wise to detour a bit and take on a front at right angles rather than slugging it out for a long period along the front.

The analysis of weather in lows and fronts tells what we may have to contend with en route; this is what we are looking for when we study the big picture.

Checking the en route weather, the VFR pilot decides if the ceilings and visibilities are high enough for him to fly safely under clouds. If he decides they are, he then asks if they will stay that way. He wants to know if any system is moving in that will make the en route weather deteriorate. If there is approaching weather but the pilot thinks he has time to make his flight before it reduces the ceilings and visibilities below a workable limit, he must decide where he will go and what he will do if the deterioration occurs ahead of schedule halfway between takeoff and landing with visibilities going down and the ceiling forcing him lower and lower.

We must remember, and it's very important, that the timing of forecasts is not always exact; they cannot hit precisely the moment when a front will pass, or a place fog in, or clear, or

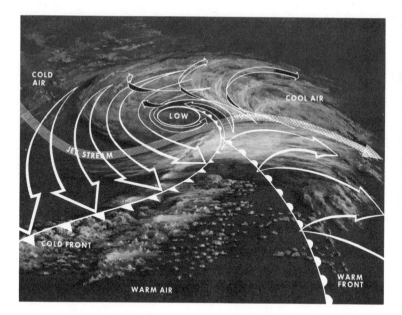

A weather system—the big picture. Note all the clouds and problems in the northeast sector. Thinking of this picture and then relating it to your location in an actual weather situation can tell you what weather to expect and which way to go. It's a picture that deserves a lot of study and mulling over. (Note the jetstream which would be up high, near the tropopause.) (NOAA PHOTO)

any other weather. The state of the art is not such that weather can be forecast, all the time, with absolute accuracy. That's why we must know weather and watch its movements all the time. I may repeat this theme many times, but I do because it is so important. We may be puzzled by weather's changes, but we should never be caught off balance!

The instrument pilot looks at the weather en route differently. He can accept weather that has ceilings and visibilities well below those necessary for safe contact flying. His question is, How tough will any turbulence, ice, or thunderstorms be in relation to his equipment and ability?

IT TAKES TIME TO KNOW

It would be easy to say that if a man is going to fly instruments he should know how to fly instruments extremely well and should have all the equipment; but this isn't very realistic. Even an FAA-rated instrument pilot has to go through a learning period. The rating, though earned after a lot of hard work, is only a beginning.

The FAA says that a rated pilot, new or old hand, may take off, go immediately on instruments, stay on instruments, fight ice and thunderstorms, shoot a 200-foot ceiling in heavy rain at the other end, miss the approach, pull up on instruments, and go to an alternate and make an approach there. That's a pretty exciting day's work for the newly rated pilot, especially when he's without copilot and busy getting ATC clearances and tuning radios.

It's obvious that the less experienced pilot doesn't want to fight all that (sometimes the experienced one doesn't either); so what does the new man do? He learns to crawl, then walk, and finally run; that is, he takes on weather a little bit at a time, gaining experience as he does.

Seeing a weather map is an important part of studying en route weather because it will graphically display fronts and systems; but often a map isn't available. Then we must get all our information from forecasts.

These start with a synopsis—which demonstrates that the Weather Bureau likes to begin as we do, by looking at the big picture. Then they go on to tell cloud bases and tops, where icing and turbulence will be, and what the outlook is for the next weather period.

We read this forecast carefully and try to visualize where all this is going on. We try to make a dull bunch of words and symbols come alive and paint a picture. It may say "C12X¾-SW 01015G25 OCNL C3X½SW," which doesn't seem to make much of a picture; but it can if we concentrate.

The ceiling is going to be 1200 feet, which isn't bad; but three-quarters of a mile visibility, variable in snow showers, gives a gloomy sight of flying in severely reduced visibility. We'll see only straight down and very little ahead. The ground will be snow-covered and all look much the same; navigation will be difficult. If the terrain is hilly with mountains nearby, we'll have to be extra careful. It's not a VFR operation.

The report continues to say that the ceiling occasionally will be 300 feet, one-half-mile visibility in snow showers. We picture sudden deterioration in conditions as the wind picks up and heavier snow showers come in gusts. Your skeptical nature tells you that the visibility will not be one half from the cockpit, but more like one quarter mile. We know it will require an instrument approach to the runway.

Probably one of the most important points in these forecasts is the outlook. Here's where we try to beat the factor of surprise. It's simply a matter of reading the outlook and then, as we fly, watching to see if the weather is working toward the conditions the outlook called for or not. If it is, fine, but if it isn't, then more careful checking is in order. Careful checking means getting the en route weather reports each hour and writing them down, in order to compare the latest with earlier ones and to see which way things are going. Don't be bashful, either, about asking by radio for the latest forecast or any supplemental ones that may have been issued.

Important in our en route flight are the winds aloft. First, of course, we study them to find their direction and velocity for flight-planning purposes, so that we know how long the flight will take and how much fuel we'll need. Add 20 percent to headwind velocities and take away 20 percent from tailwinds as a "fudge factor." That's about as accurately as they are forecast on an average.

The winds should be studied at various altitudes. We are generally accustomed to the idea that the winds are stronger the higher we go, but this is not always the case; sometimes we find lower velocities up higher—which could help if we were flying into a headwind. We study the wind at various levels in order to choose the best altitude at which to fly. An old rule we used in DC3 days when flying into headwinds was to fly high in southwest winds and low in northwest to reduce the wind effect. It's not always true, but it's not far off.

Wind is really a secondary factor in selecting altitudes; weather is first. We wouldn't want to fly in a deck of icing clouds merely in order to have a better tailwind or less headwind. So weather is really the first factor in picking an altitude (besides terrain clearance, of course, which comes before everything), and wind is second. Airplane performance is another factor. We want to cruise as close to the airplane's optimum cruise altitude as possible; weather and wind will dictate whether we can or not.

It's good to keep in mind the wind at other levels, in case ATC sends you to a different altitude than the one you want. The trip might take longer at a different level. How will this affect fuel reserves? In a jet this could be quite serious, but it's important in a 100-horsepower light plane, too.

The winds aloft also give weather clues. Suppose the wind for the trip to be forecast southwest, but as you fly you find the wind really coming from the southeast. That's a clue that the weather may be starting to do something different from what was forecast, and it's time, again, to watch developments more carefully.

A last important point about studying en route weather regards flying around weather. We should not look only at the weather on the direct line between where we are and where we want to go. The old idea that the shortest distance between two points is a straight line doesn't always apply in flying. We should look well to either side of our route and see what's going on. Sometimes we can circumnavigate bad weather. Sometimes making an end run is worth it just to get better winds, because we are not interested in how far we fly in an airplane, but how long it takes. Going "off course" to get better winds may actually make the flight shorter.

Ducking weather, in a big sense, if we're flying from the Midwest to California in winter, it's generally better to go via El Paso than via Denver; the mountains are lower, the weather less violent.

From Columbus, Ohio, to the East Coast in winter, it may be cleverer to fly a southern route—say, Morgantown–Martinsburg, West Virginia–Westminster, Maryland—rather than direct because the more southern track would keep us away from the snow showers and clouds that are generated by the Great Lakes.

Even short flights sometimes can be flown in better weather by a small course deviation. It pays, in looking at en route weather, to look not only up and down, but on both sides too.

6

Equipment Needs for Weather Flying

IT IS ALMOST IMPOSSIBLE to talk about flying weather without talking about the equipment of both the airplane and the pilot.

The airplane we fly must fill certain requirements, not only of instruments and radio, but also of power and range to battle the elements when necessary. This doesn't mean it has to be a big four-engined transport, or even a small two-engined one; but the airplane should have a decent rate of climb, and more important, it must have enough range for the flight, from takeoff to destination, to be made against a strong headwind and along the wiggling course that may be necessary because of ATC. There is an impressive difference between the straight-line distance between two places and the distance measured over the wandering routes of the Federal Airways System.

IT'S FARTHER THAN YOU THINK

While the shortest distance between two points is a straight line, airplanes, flying instruments over the airways, don't fly that way. One flight I often make is a 251-mile direct flight from Van Sants, a small field in Pennsylvania, to Montpelier, Vermont. If I file instruments, however, the airways distance is 304 miles—that's over 50 miles extra, and in a Skyhawk that's a lot.

A flight made around the New York area was 60 percent farther by ATC cleared route than over a more direct filed route that also avoided the New York airports area. The big extra distance takes lots of additional fuel, and that affects safety. Suppose we were trying to beat darkness and approaching weather—a slightly stupid combination—and then found our ATC clearance via a route, such as this, that would take seventy minutes longer, not counting possible headwinds. We could easily run low on fuel, fly into darkness, and meet bad weather head on!

If there's any doubt, check the preferred route with ATC in advance via FSS or telephone to the Center.

Range also must include enough fuel for reasonable holding at the destination and then diversion to an alternate airport against a headwind. Furthermore, of course, some fuel should remain in the tanks when the pilot arrives over his alternate.

This is the needed range:

1. Departure to destination, plus
2. Holding fuel, plus
3. Fuel to alternate, plus
4. Reserve over the alternate.

Lets break these down further.

1. *Departure to destination,* the long way. It is most important to know the en route time, with winds taken into consideration. If we are going through bad weather, the wind may be strong. Slugging against the winds of an active low-pressure area may give ground speeds low enough to cause apprehension.

We need fuel for departure and arrival wanderings as directed by radar. You do not simply fly to the destination's radio aid, make an instrument approach, and land. If it's a busy terminal you may be told to fly vectors that add many miles to the distance already flown over the wandering airways. One route into JFK from the north has an average vector distance of 91 miles! If we're flying a jet, and doing it at a low altitude which we may have been cleared to, those 91 miles can use up a lot of fuel and make our reserve a considerably, nervous, lower amount than we'd like to have. (Sometimes these vectors carry one over big chunks of water in the vicinity of coastal or lakeside cities; this makes buoyant cushions that can be used as life preservers a good piece of airplane equipment.)

Fuel consumption is higher and speed lower during climb; so the altitude to be flown will affect the total fuel needed. This can be determined with a knowledge of the average rate of climb, the average true airspeed during climb, and the fuel consumption at climb power. A pilot can make a handy chart that will give the fuel burned and the miles made good by his airplane in climbing to various altitudes as a ready reference in flight planning.

WEATHER USES FUEL

In icing conditions the airplane can lose speed, and because of extra power and/or carburetor heat, the fuel consumption will go up.

Wandering around thunderstorms also adds miles and consumes additional fuel.

It is difficult to say with precision how much to add for ice or detours around thunderstorms, and pilot judgment must prevail. But awareness of the fact that these things take fuel is a good beginning, and a 10 percent cushion will serve as a start.

2. *Holding fuel.* Again it is difficult to judge how much we need, but there are factors to help us make an educated guess. First, is our destination a busy terminal? If so, the delays can be long. Obviously if we are going to Chicago's O'Hare, the delay will be longer than for Reading, Pennsylvania.

   Time of arrival may also affect holding time. Any big terminal will have much longer delays between 4 and 8 P.M. than at 1 P.M. or midnight.

   Holding ability depends also on the airplane. A piston-engine airplane can get its fuel consumption down quite low when just lolling about. A jet, however, uses lots of fuel and is especially critical at low levels. If jets can hold above 20,000 feet, their fuel consumption can be brought within reasonable limits. Fanjets are less greedy than straight jets, however, especially at low levels.

   The weather forecast enters into the holding problem. A person doesn't want to hold, for example, while waiting for the arrival of a cold front. If ATC says the hold will be for an hour and a cold front is due within that hour and if reports indicate that the forecast is accurate, it may be wise to proceed immediately to the alternate and forget about holding. If it is late and fog has formed, there is no point whatever in sticking around. Which, again, emphasizes the importance of keeping up with a range of weather reports in flight.

As a yardstick for the *minimum* holding fuel, there should be enough for one hour and a half at a busy terminal and 30 minutes at more out of the way places.

Traffic can back up badly at a busy terminal or alternate because of thunderstorms, during which landings may cease entirely. Traffic keeps pouring into the area, and stacks up even worse than in winter conditions. I've seen Kennedy go from 15 minute delays to two and a half hours in a matter of minutes because a thunderstorm hit the field—*it* had number one priority!

3. *Fuel to alternate* is just like fuel to destination and should include the same factors—i.e., total "long" distance, wind, and weather en route. If there is not a headwind to the alternate, the fuel consumption can be computed on a long-range cruising basis, and thus less will be required. But if there's a headwind, long-range cruise doesn't help enough to bother with and if the headwind is strong, it will not help at all. This is worth analyzing by doing some calculator or computer work at leisure on the ground.

Headwinds affect airplanes according to their true air speeds. A 50-kt headwind against the Concorde is about 3 percent, or something like six minutes extra time on a Paris–New York flight of 3200 nautical miles. But 50 kts on the nose of a Cessna 172 is 45 percent, which means an hour and nine minutes extra between New York and Washington, about 175 nm!

It's important to explore head- and tailwind numbers and to know what they do to range and fuel on our individual airplanes at normal and long-range air speeds and different altitudes. With this knowledge we can quickly set up for maximum range when we really need it.

4. *Reserve over the alternate* depends on weather at the alternate. If it's clear and forecast to stay that way, the reserve can be minimal. If the alternate isn't clear and instrument approaches are required, even with a fairly high ceiling of 1500 feet or so, more fuel is needed because there may be traffic delays there too. Often, when an airport closes down or becomes unusable, everyone holding at that airport flies off to the same alternate. This moves the congestion from one airport to another, and the alternate can be a madhouse of delay and confusion. Arriving over the alternate with minimum fuel in this situation is a scary proposition.

As we begin weather flying, of course, we should pick alternates that are clear and definitely forecast to stay that way. When we say "clear," we mean broken clouds or better and ceiling 2000 feet above the highest terrain within 50 miles.

FUEL AND THE LAW

Alternate fuel reserve should be 45 minutes—the Federal Aviation Regulations say at normal cruise power setting—but it's good to keep in mind that this is the final fuel, and when the last of it slides through the engine, all problems become simple. You are going to land, right now, wherever you are!

The government regulations spell out fuel reserves, but these certainly should be considered minimums. This is all they ever could be, because it's impossible to write regulations that cover all conditions—especially considering the fickleness of weather. Having sufficient fuel is one of the greatest safety factors one can give oneself. It assures the mental tranquility that is of paramount importance in weather flying. A pilot running low on fuel may force himself into a hurried emotional decision that is wrong.

It's worth reflecting that fuel management is a major cause of engine failure. Good management means, for one thing, having enough—not running out! It also means having the good sense to be certain the fuel valve is always on a tank with fuel in it. Airplanes have run out of fuel and made forced landings with the fuel valve on an empty tank, while fuel remained in another tank!

The added advantage of being able to accept a wandering route from ATC, without being unnecessarily concerned about fuel, makes weather flying an interesting, enjoyable experience rather than a nervous and jittery ordeal. The ability to fly out of the weather area and go where it's clear is a solid comfort of the first order. A pilot can never allow himself to be in a position where all his bridges are burned and there is no way out. A fat fuel reserve goes a long way to provide the necessary out. Desperation due to a dwindling fuel supply has undoubtedly caused more weather accidents, by far, than engine failures.

FUEL AGAIN

There are two ways to deal with fuel. One is to have lots of it—and airplanes are getting better every day in this regard. Certainly if maximum utility is desired, an airplane's manufacturer should strive to provide for the greatest possible fuel capacity when the airplane is originally designed.

The second is to fly within the airplane's fuel capacity by limiting the length of flights. This is restrictive, of course, because sometimes weather covers a big area that doesn't allow a short-range airplane any alternates. Then there's only one thing to do: Sit and wait for better weather.

INSTRUMENTS

What else does the airplane need? Instruments to fly by, arranged in a good useful fashion. The hit-or-miss manner in which instruments are often stuck about cockpits is appalling. Good instrument flying requires constant visual scanning by the pilot. Logically, then, the shorter the distance a pair of eyes needs to travel to scan an instrument panel, the better.

Each panel arrangement presents its own problems. The space available, control-wheel location, windshield height, and other such factors dictate how instruments can be arranged; but regardless of these problems the flight information instruments need to be grouped close together.

The basic key of instrument flying is to keep one's eyes constantly roving over the important flight instruments. When a pilot flies weather, the technique of flying by instruments should be so well developed that he can fly easily and devote his attention to the problems of weather, air traffic, and flight management. Ideally, an automatic pilot does the job, but not all airplanes are equipped with them, and generally during low approach and takeoff the pilot is not using the autopilot.

Autopilots have reached a level of sophistication that benefits all aircraft tremendously. With modern big aircraft, autopilots can make approaches, kick out crab, land, and roll the aircraft to a stop, as well as do all the flight chores such as climb, descend, and follow navigation signals, be they simple omni, advanced Omega, or Inertial. Smaller single- and twin-engine piston aircraft have autopilots that follow headings, omni or ILS, and hold altitude, plus other good things. These are light and reliable because of transistors and microprossors.

In the complex world of instrument flight an autopilot is almost a requirement—someday they will probably be standard on all except pure sport airplanes; they are certainly very desirable, as they turn a busy hand-flying task into a relaxed experience where the pilot is well in command of the situation and always ahead of the airplane. The lower a pilot's experience level the more he needs an autopilot, or, of course, a good copilot.

The key instruments are the Artificial Horizon, for bank and pitch reference, and the Directional Gyro for heading. The Directional Gyro is fast being replaced by the Horizontal Situation Indicator (HSI), which is a big advance and help because it has heading and navigational information all together; the heading is slaved to a remote pickup and stabilized so one doesn't have to constantly worry about resetting the DG. Airspeed, Rate of Climb, and Altimeter are really reference instruments, and their action is a result of things that appear first on the Horizon and Directional Gyro.

If the key instruments are watched closely, the airplane never gets a chance to go very far from the "straight and narrow." Frequent scanning and immediate correction of excursions from the desired flight path give the feeling that one has the airplane in a narrow corridor, boxed in on all sides, unable to escape and fly off on its own.

Since rapid scanning prevents the airplane from getting far off course, any correction the pilot needs to make will be small and therefore easy. Problems start when a big bank or pitch angle is allowed to develop. If this happens, the heading, rate of climb or descent, airspeed, and altitude go off, and the pilot has his hands full trying to get back where he belongs. Scanning is not difficult if practiced. Small corrections and never allowing the airplane to wander far make flying simple and relaxed.

Small simulators, such as the desk top types available, are excellent trainers to improve one's scan and keep it up to speed. They improve our ability to handle complex ATC procedures and are an answer to the question of how to stay proficient when the autopilot is doing all the work.

It is very important that the pilot be well versed in attitude flying. There are reams of information and instruction available, and an instrument pilot should take advantage of them to become a good attitude pilot. Attitude flying, in one simple example, is the difference between attempting to maintain a specific airspeed by chasing the airspeed indicator, rather than by keeping the Horizon bar in a position that will give the specific airspeed desired. If speed is too high, the nose is raised slightly with reference to the horizon, thus producing a small but positive airspeed correction. Practice soon teaches what attitudes will provide the desired airspeeds, heading changes, descents, and climbs in small, easily controlled increments. But more important, it will help keep airplane movements small and prevent large oscillations that are difficult to recover from.

Well-arranged instruments, combined with good scanning and attitude flying, make instrument flight precise and simple.

EQUIPMENT NEEDS

There are some other things airplanes should have when flying instruments, and we list them here.

PITOT HEAT   Airlines feel pitot heat is so important that they use it whenever they are flying, cloud or no cloud, summer or winter, to be certain it's always on in case of need. A temperature drop can take place inside a pitot tube and produce ice or slush if there's visible moisture. Even water from heavy rain can cause an erroneous reading in pressure instruments. Ice, of course, can knock out an air-speed indicator completely. Pitot heat is a must!

WE CAN KEEP IT SIMPLE

What kind of instruments does one need? The extent and sophistication of the instruments is determined by the amount and kind of weather a person is going to fly. A pilot can fly some pretty awful weather with quite simple instruments; but he's very busy, and the job accomplished won't be a neat, precise one.

My own instrument flying began with an Airspeed, Vertical Speed, and a Turn and Bank. I taught myself how to use them by following a book called *The 1-2-3 System of Blind Flying,* which was written by an airmail pilot named Howard Stark, one of the important pioneers of instrument flying.

I flew actual instruments, too, but it was all restricted to climbing up through or flying on instruments toward clearing weather. It was done without radio, by simply holding a heading. This wouldn't be possible today in most parts of the world because of airways and traffic; but back then there wasn't any traffic, nor any airways either.

A LITTLE MORE TO DO A LOT

From that simplicity I graduated to a Douglas DC2; the only addition to its instrument panel over my little Monocoupe's was an Artificial Horizon and a Directional Gyro. Along with this, there was, of course, a radio to follow beams and a loop turned by hand to read bearings which said the station was at one end of the bearing or the other, but not at which one, because the loop didn't have direction-sensing capability.

With this setup we flew weather, lots of weather. The landing minimums at Newark, New Jersey, in 1937, were 200 feet ceiling and one-half-mile visibility! Takeoff minimums were 100 feet and one-quarter mile! It wasn't done very sophisticatedly either. We flew the radio beam toward the station which was down in the Newark meadows near Elizabeth, New Jersey. We crossed over the center of the station, appropriately called the "cone of confusion," at 800 feet, chopped the throttles, shoved the nose down, and descended as fast as possible to 200 feet; then we looked into the black night for a row of red neon lights that led to the black runway. What's so surprising, in restrospect, is that we did it often and successfully. Of course, a DC2 could fly at 80 miles an hour; that makes a big difference from a jet that you don't get under 135 knots.

All this is to show that a lot of weather can be flown with a primary flight group, an Artificial Horizon, and a Directional Gyro.

Our old Artificial Horizons did about what today's can. The Directional Gyro was the bore, because like DGs of today it didn't have any north-seeking ability and needed to be set to agree with the compass. Gyros precess and must be set frequently. They must be set when the airplane is level and not turning, and, of course, the compass must be settled down so that it's reading accurately. In rough air, while working on omni or ILS, this can be tough. Some gyros precess more than others—depending, generally, on the condition of the gyro. If it is well maintained and periodically overhauled and if the vacuum source, in the case of vacuum instruments, is set to give a constant value of vacuum, a gyro can hold a heading for 15 minutes or more without much attention. A point to remember is that once a gyro starts to precess, the rate of precession increases the more it precesses; so it's wise to reset a gyro before it gets too far off.

THINGS CAN BE BETTER

In modern instrument flying, which requires precise following of airways, a DG may add appreciably to the cockpit work load. Instrument designers have cured the problem by wedding the Directional Gyro and the compass so that a north-seeking unit constantly keeps the gyro's heading adjusted for magnetic north. This is done in different ways by different manufacturers and given different trade names; but the important point is that once the gyro is set the pilot need never worry about resetting it, and he has precise heading information before him at all times, without the turning error, run ahead, and hold back of a magnetic compass.

This intelligent type of gyro has been one of the biggest boons to precise instrument flying, and it reached aviation technology at about the time that heavy traffic and more narrow airways called for better flying. This, now, is the HSI we talked about; it's one of the first "extra" items I would put on my instrument panel.

To sum up at this point: The minimum one needs for instrument flight is the primary flight group plus an Artificial Horizon and Directional Gyro. With this, one can fly instruments and shoot low approaches; with the addition of a slave gyro or HSI, however, life becomes easier—and *much* easier with an autopilot.

EVEN BETTER

Technology goes further. It has added intelligence to the other instruments. To explain a bit, if we fly an Instrument Landing System, ILS, with a Horizon and DG, we fly headings toward the localizer; once on it we try to find a heading that will keep us on it, by a process called bracketing. How much to turn and when to turn is a matter of skill and, to some extent, guesswork. As Dave Little, an American Airlines captain who has done a great amount of research in instrument flying, said, "Flying down a localizer this way is like following the white line on a road by watching it through a hole in the automobile's floor."

But now we have Artificial Horizons that have computed information fed into them. They discover how far off course one is, what drift there is, when an airplane is wandering from the course, and then say, "turn now and this much." The information is presented in such a way that one has only to follow a little airplane, or bar symbols, superimposed on the horizon, matching one with the other, and by so doing a pilot stays right on localizer and glide path. He still refers, although only periodically, to the basic localizer/glide-slope instrument to make certain that the flight director, as some of these instruments are called, is doing its job.

With equipment of this sophistication aboard, jets are allowed to shoot minimums as low as 200 feet. The flight director is required for ceilings of 200 feet and the autopilot below that. If the flight director is not working, minimums are higher; this is one reason why airliners carry duplicate systems. With this type of equipment a 200-foot ceiling is easier to fly than a 400-foot one without it.

The computer data fed into these flight director systems are the same as that given to the automatic pilot when it is set up for making low approaches; if the airplane is equipped with both (as airliners are), one system is used as a check against the other.

With the Boeing 747 came autoland, and our minimums were reduced at certain airports to 1200 feet visibility and 100-foot ceiling. The many approaches I made to these low minimums were easier, more relaxed operations than flying by hand to a much higher minimum. The autopilot does the work and you have lots of time to scan the cockpit and double-check instruments. You have an understanding of the entire action rather than being mesmerized by the few instruments that one uses to keep in the slot when flying by hand. When the lights come into view, there's no worry about sensory illusions and getting too low and below glide slope on the visual part of the approach because you're there! Right at the runway—and in a few seconds you're rolling down the center line on the ground. It's the way to do it.

THE FUTURE WILL BE EVEN BETTER

The future promises even more useful instruments, such as ones that tell us precisely and quickly what power is needed for each condition, airspeed indicators that say at what rate we are increasing or losing airspeed, and a host of other innovations in presentation as well as information. With this equipment we will fly lower and lower minimums and some day land zero-zero; but the fellow with a Turn and Bank, Airspeed, and Vertical Speed indicator will still be able to fly instruments. He will be limited to the amount of instrument flying he can take on, just as the pilot with a little more equipment is limited, though not quite so much, and the man with everything is limited even less.

The equipment, humble or extensive, must be used properly, with the pilot's proficiency honed to a fine edge through both knowledge and practice.

It is a good idea, too, for the man with the fancy equipment to maintain a certain proficiency with the simpler equipment. The basic flight group, Airspeed, Vertical Speed, and Turn and Bank, are the most reliable, and they are always standbys in case other instruments fail. For more than 40 years the Turn and Bank has been found on every instrument panel with which any instrument flying was done, from Cubs up to airliners and jet fighters. The Turn and Bank is beginning to disappear from airliners and some large aircraft, being replaced by a small standby Horizon. But most General Aviation airplanes still have it; it's a good instrument and keeps working no matter what the aircraft's attitude is. If one is on your instrument panel, it's a good idea to cover the Horizon and DG now and then and have a practice session with the Turn and Bank only. This is good simulator exercise too. Many Turn and Banks are no longer needle displays, but look like an Artificial Horizon and are called Turn Coordinators. Remember they are not Artificial Horizons, but gimmicked-up Turn Indicators, and should be flown as such. They still retain the valuable property of working regardless of attitude, however.

POWER FOR INSTRUMENTS

Instruments, however, are no better than their power source. Some are vacuum-powered and others electrical. There should be an alternate source for gyro instruments in case the basic source fails. This can be supplied in a number of ways. If the instruments are vacuum-powered, then there should be two vacuum pumps on twin-engined aircraft, or a standby Artificial Horizon or Turn and Bank run by batteries, such as we have in sailplanes. A Turn and Bank can be powered by an outside venturi. Although people shudder to think of venturis sticking out causing drag and getting iced up, a system I had on a Pitcairn Mailwing, vintage 1930, took care of at least the latter problem well. The T&B venturi was mounted on an exhaust manifold with five tiny holes in the manifold just ahead of the venturi's entrance. These squirted hot exhaust gas over the venturi, and prevented it from icing up. Such a rig might not be a bad alternate source today.

Most modern instruments are electrically powered. It is important to have, for these, an electrical system with two generators. Most batteries are just a wide place in an electrical wire and will not supply all the airplane's electrical needs very long on their own.

There have been some dramatic situations and accidents because electrical systems lacked sufficient backup. If an airplane is going to be used for instrument flying, its electrical system should have a backup.

There are some wonderful solid-state radios that run from dry batteries and do so for a long time. My glider has a 360-channel two-way radio with good quality and power. I've run it over 125 hours on two small 7½-volt batteries. One like that would be a big comfort in an airplane. It always would be there, independent of the electrical system.

The Airspeed Indicator and Altimeter are "powered" by pitot and static tubes that bring in pressure and static air. Since it's very difficult to fly without airspeed, it's important to keep these sources clean and operating. We've already mentioned pitot heat. If the static source is within the pitot head, then the pitot heat will take care of it; if the static source is flush on the outside of the airplane, as most are today, it will not be closed by ice, but it can be blocked by slush thrown up from a nosewheel and then freezing, or by man-made things like tape put over the static holes during airplane washing or servicing. Careful inspection is necessary before flight to be certain that the static source is free and that the pitot tube is also. A bad static source will affect the Altimeter, Vertical Speed, and the Airspeed.

The FAA requires that the condition of these instruments should be checked periodically for leaks and deterioration of the tubes in the system. Instrument flying is important in weather flying, and while a person may or may not be proficient, all pilots can make certain the instruments they are using will be in top shape and functioning properly.

INSTRUMENT LOCATION

Almost as important as the condition of instruments is their location. The basic T concept, developed many years ago by the Air Line Pilots Association members is still best. What it does is put the most needed and worked instruments in front of the pilot. The most important, of course, is the Artificial Horizon. We concentrate on this instrument most, and it should be directly in front of the pilot. Beneath it should be Heading; the Artificial Horizon and Directional Gyro are the basic, central instruments. Ideally, the ILS-Omni indicator should be close under the DG (by DG we mean the Directional Gyro or any sophistication of it), so that quick reference between the two can be made during tight ILS approaches. In the most sophisticated instruments ILS and Heading are in one instrument, the HSI, along with the Flight Director, as part of the Artificial Horizon. But whether instrumentation is simple or complex, the Horizon, DG, and ILS-Omni indication should be vertically stacked if at all possible.

Off to the sides, but close by, are the Airspeed, Altimeter, and Vertical Speed. These are referenced against the Horizon and DG and should be vertically stacked if at all possible.

The important point is that the instruments should be as nearly in front of the pilot as possible, and they should be in clear view and not hidden by the control wheel or anything else. Pilots who fly one of our most used jet airliners have suffered because the Heading information and ILS-Omni instrument is partially hidden by the center of the control wheel, and the only way they can adequately see the instrument is by sitting too high and too close to the wheel. The pilots say that the only way to get the seat adjusted properly for instrument view is to move it up and forward until it's uncomfortable! Too bad.

LIGHTED WELL

It is important to have the instruments lighted properly. This means that at night the entire area of their dials should be illuminated without any shadows. There should be no glare, and they should be visible in a minimum of light. It should not be necessary to turn the lights up bright or to have excessively bright lights in order to see the instruments. Some airplanes I've seen require such bright panel lights to wipe out shadows on the instruments that outside vision is impaired. The pilot is like an actor trying to see the audience through bright footlights. The bright light not only cuts out what can be seen outside, but deteriorates a pilot's night vision. Bright lights also have a way of reflecting in windshields, so that as a pilot looks outside, he sees instrument dials grinning at him from the windshield or side windows.

On the subject of night flying, it's worth realizing that night VFR flying is instrument flying to a large degree; while a noninstrument pilot can fly at night, he's going to need a clear night with good visibility and some sort of horizon. It can be an awful shock for a pilot to first fling himself into a pitch-black night to find he doesn't have a good horizon to fly by, and then have his hands full keeping control.

If the horizon is not clearly defined and there is any risk of clouds, it's easy to slip into a cloud and never realize it until all outside reference is gone and we're on instruments! An instrument rating is a strong safeguard.

PAPERWORK IS EQUIPMENT, TOO

Almost as important as an airplane's instruments is the paperwork associated with instrument flying. We need maps, radio facility charts, and instrument approach plates, and they must be up-to-date.

But besides having up-to-date charts, plates, etc., we have to have a place for them in the airplane. There's nothing worse than charts scattered all over the cockpit.

Good instrument flying is good housekeeping. This means a little preparation in advance, before takeoff. The pilot ought to get out the charts he'll need for the flight, stack them in order, put them on a clipboard or some such device, and keep them close by so they are handy when he needs them.

Instrument plates should be treated the same way, and a clever pilot will have a clip or some such gadget on the control column where he can fasten the plate for reference while making an approach. It should be lighted.

The charts that we fly with are confusing and often difficult to read and interpret. There have been serious mistakes made because of the vague way in which things are sometimes presented. The papers we work from are a built-in hazard and should be treated as such. It is necessary to study routes and terminals in advance so you don't have to do it in the cockpit under poor light while bouncing around and trying to fly the airplane.

It's a good plan to have a systematic procedure for studying instrument plates. I always start with the field elevation, then the minimums, the ILS course, frequency and identification, the altitude of the glide slope at the outer marker, the minimum altitude I'll go to before pullout, then what the pullout procedure is in case of a missed approach. I look for the highest obstruction on the charts and note the minimum safe altitude within 50 miles. Then the runway length, lights, and approach lights. Also very important are the little notes that chartmakers stick off in corners that say things like, "Do not descend below 4400 feet on approach to runway 21 until Radar Controller advises passing 4074 feet stack 5 NM NE of apt."

Now we have Standard Instrument Departures (SIDs). They create another chart that must be consulted before takeoff. It is important not to be rushed into taking off until one understands what the SID says and has it firmly implanted in his mind.

Similar to SIDs are STARs (Standard Terminal Arrival Routes), except, of course, the STAR routes are for arrivals and show the proper routing, altitudes, and sometimes speeds. Combined with STARs are Profile Descents for high-flying aircraft. Sometimes Profile Descents have their own special charts too.

All these are important because there is lots of "fine print" which makes it easy to miss some ambiguously worded instruction. It's best to study these carefully in advance, on the ground and at leisure, and then to refresh one's memory before starting descent, preferably in a relaxed, cruise portion of the flight.

Felt marking pens to overlay routes, altitudes, or any other important bits of information are a big help. Yellow is a popular color, but almost any color allows one to see immediately the area being used rather than hunt all over the chart for it when busy in flight.

GO FAST SLOWLY

There's a great tendency at busy airports for clearances to be delivered too fast and followed by a rushed "Take position, be ready for an immediate takeoff!" It comes in firehose fashion and tends to make people take off half understanding how and where to go when in the air.

The answer is simple: to heck with the tower. Sit there and study the clearance. If you don't get it, ask for a repeat. Then, when certain, ask for takeoff clearance. It's better to delay the takeoff considerably rather than rush off half-informed.

In the cockpit we need a place to write clearances, then checkpoint times and estimates. We want to copy weather and clearance changes, keep track of fuel used, and write down a lot of other things. We need a clipboard or kneeboard strapped on, handy for the job.

A computer and plotter, big enough to see and handy at all times, are also indispensable pieces of equipment.

GOOD HOUSEKEEPING

All this is part of instrument flying, and keeping a neat, organized cockpit makes instrument flying much easier. But more important is the fact that a disorganized cockpit makes instrument and weather flying much more difficult and increases the possibility of serious mistakes.

We like to have a copilot to help reduce this instrument-flying load; but even if we don't have an official copilot (that is, a pilot with ratings), we can often train a nonflying person to look things up and sometimes even to handle the radio. This is best done by sessions at home when the pilot goes over the airways, approach, and navigation charts with the nonpilot and explains what they are, how certain ones are needed in flight, how to find them, how to add mileages, how to fold charts for pilot convenience and point to where they are, how to dig out frequencies, and many other small but useful duties. This can expand to learning to use a computer for speeds and fuel, and to limited radio work. A radio license would be required, but they are simple to get.

A little training of this type may turn wife, child, or friend from a bored or nervous passenger into a useful, interested, and relaxed crew member, even if they don't know how to fly.

BOOM MIKE

Without a helper the work load gets pretty heavy. One way to reduce it is by using a boom mike. The idea of fumbling around to find the mike, getting it off the hook, holding it up and pushing the button, and taking one hand out of use while doing it is terribly primitive. With a boom mike and a button on the control wheel one need only squeeze a finger to communicate. The other hand is free to write messages, adjust gadgets, and do the many chores it may be called upon to do. Boom mikes are thought important enough to be required by regulation on transport aircraft in the United Kingdom and the FAA requires it on the single-pilot Cessna Citation. An excellent idea.

7

Temperature, an Important Part of Weather Flying

TEMPERATURE is closely related to many things we do with airplanes. It affects performance in takeoff, climb, cruise, and landing. It's important.

Air is the thing our airplane flies by. How much or how little air relates to the number of molecules. The amount of air striking the wing is a function of speed and of the number of air molecules present. If cold, the molecules are packed closely together, and we say the air is dense. If it's hot, there are fewer molecules per cubic foot, and the air is less dense. We sometimes say it's "thin."

The more dense the air is, the better our peak performance. Engines put out more power; the wing has more lift in dense air. When the air gets hot and thin, it's all reversed and performance suffers.

TEMPERATURE AND DENSITY

There's another way of looking at it. At sea level the air has a certain density, but at altitude the density is less. We know how this affects performance. When we take off and start climbing, our rate of climb is, say, 1000 feet per minute. As we climb, the rate decreases until at the airplane's absolute ceiling there isn't any rate of climb.

Why did the airplane stop climbing? It's simple: it ran out of air. The air wasn't dense enough to feed the engine and keep its power so it could push the wings fast enough to fly in the high and/or hot, less dense air. The airplane's performance is limited by thrust, or lack of it. If there isn't enough push, or pull, the airplane stops climbing. An unwary pilot may keep pulling the nose up, trying to eke out a higher altitude or stay at a marginal one, and get to an angle of attack pretty near stall—and that's a fragile situation! So without air an engine doesn't do its job, wings don't lift, and humans don't live.

This happens as we climb because the air is less and less dense at higher and higher altitudes—not because of heat or cold, but because the atmospheric pressure is lower. The higher we climb, the less air there is above us pushing down and creating pressure. It's always impressive to realize that half of the atmosphere is within the first 18,000 feet of altitude! At sea level the atmospheric pressure is about 30 inches of mercury. At 18,000 feet it's about 15 inches of mercury.

Since hot air is less dense than cold air, hot air acts like air at high altitude, and cold air acts like air at lower altitude. What we've said is that air density and effective altitude are related. A sea level airport on a hot day isn't at sea level at all as far as our airplane's performance is concerned. How high sea level can be sometimes comes as a surprise.

A sea level airport on a day in summer, when the temperature is 38 degrees C (101 degrees F), has a density altitude of 3000 feet! The sea level performance of the airplane will have deteriorated about 20 percent. This figure will vary, especially with supercharged engines. In any case the airplane must not be overloaded; it must be flown carefully, and there must be enough runway for safe operation.

USE THE COMPUTER

The density altitude can be found on any good pocket speed-distance computer. Most of them include a density-altitude computation window. It's worth a winter's evening to run out some make-believe situations and see how density altitudes can vary.

Airplane manuals generally show performance at different altitudes. The density altitudes we find on the computer can be compared with the airplane's performance at that altitude to see what performance we can safely expect on hot summer days.

HOW HOT, HOW HIGH?

If one flies out of high-altitude airports, 5000 feet or more, with hot temperatures, it's not unusual to have a density altitude of over 8000 feet! Some low-powered airplanes will hardly fly at that altitude, and even the best will have impressively reduced performance.

Remember, standard temperature at sea level is 15 degrees C (59 degrees F). That's what all those manual figures are based on. The moment the temperature has gone one degree above 15 degrees C, the performance has started to deteriorate.

ENGINES DON'T LIKE IT HOT

Heat affects the engine's running too. It doesn't put out as much power because it is flying, in effect, at a "higher" altitude.

Running on the ground can heat up heads and other engine parts excessively. Tightly cowled engines need a lot of cooling airflow, and when we are sitting on the ground in hot weather, the engine isn't getting the flow it may need. We should keep ground-running to a minimum, and if we must sit there with the engine running, we should be certain our nose is headed into the wind to gather all the air possible.

Vapor locks occur quite easily in hot weather. Most engine systems are designed to prevent these; they aren't much of a worry with modern aircraft, but if fuel booster pumps are required to be on for takeoff, it's worse to forget them on hot days when vapor locks could occur. If we use that checklist, this will not be a problem.

All this suggests that cold days are wonderful. They are, for airplane and engine performance. The airplane leaps off the ground and it climbs like a homesick angel; it's one compensation for freezing.

But cold does create a hazard in relation to air density. It affects the altimeter. In very cold air the altimeter will read higher than the airplane's actual altitude.

For example, flying at 10,000 feet with a temperature of −32 degrees C (−25 degrees F), which is 27 degrees C (49 degrees F) below standard, if the altimeter setting was furnished by a sea level station the altimeter will say 10,000 feet, but our actual altitude will only be 9000 feet! So if we were poking around mountains counting on 1000 foot clearance, we might not have it.

Imagine a low approach to a sea level airport with a very substandard temperature of −26 degrees C (−15 degrees F), such as one can get in the north. When our altimeter said 300 feet, we'd be at only *258 feet*. It's not a big difference, but big enough when you're making an approach to a 300-foot-minimum airport.

A good rule to remember is that for every 20 degrees F (11 degrees C) the temperature varies from standard, there is a 4 percent error in altitude: lower temperature, lower altitude than the one indicated on the altimeter.

So cold temperatures are fun to fly in; but they can give us erroneous altitude information, and that's disastrous.

What this all says is that temperature, like wind, is one of the elements that a good airman becomes aware of instinctively. He checks constantly, in his mind, what the current temperature is and what it's doing to his flying.

8

Some Psychology of Weather Flying

BEFORE GETTING INTO weather flying we should talk about emotions.

Basically, airplanes get us into places and situations man wasn't designed for. The inside of a thunderstorm, for example, where the airplane is buffeted around by turbulence, where there are loud frightening noises caused by heavy rain and sometimes hail beating on the airplane and our nerves. Lightning flashes, thunder is often heard, and now and then a flash of lightning leaves an odor like the burning insulation of an electric wire.

In clouds, with ice collecting on the airplane and a very low ceiling at our destination, we can feel pretty lonely. With a little imagination a pilot may wonder if that cozy world is still down there at all.

Flying weather at times requires a firm grip on one's emotions. Emotions have to be subdued by realizing that to lick weather one needs logical, step-by-step thinking and intelligent use of equipment. Wishing will not make a successful descent through the clouds and landing on the airport. A good job of flying will.

Sometimes a pilot has to take a firm grip on his emotions and force himself to control his nervous stomach and shaking hand. This isn't always easy, and anyone who says he's never been scared flying in weather either isn't telling the truth or hasn't flown in any!

What we have to do is be big grown-up people and keep everything under control. One can be anything from slightly to completely nervous and still think; and think we must.

A difficult factor is that sometimes in flying we can be nervous or scared for a long period of time. We may sit on instruments a number of hours with everything under us below landing limits and our destination on the ragged edge too. This type of operation is known as "sweating it out," and a more perfect description was never thought up for anything.

SELF-DISCIPLINE

When everything inside us is scared we have to work harder to do a good precise job of flying, thinking scientifically all the time. In this situation a pilot must do his utmost to be relaxed. Being relaxed creates better flying and better thinking and reduces fatigue. So even if hell's fire and brimstone are all around, we must keep reminding ourselves to sit back comfortably, relax those white knuckles on the control wheel, and think! It isn't easy, but it's possible; and working on it, forcing oneself, and practicing make it possible. Surprisingly enough, effort in this can develop a control over our emotions that favorably affects the nonflying part of our lives, from driving automobiles with their dangers to personal problems of finance or heart.

There's a natural tendency for fright to speed things up. If a person is in a jam in the air his basic human desire is to be back on terra firma. If he panics, he may try to get on the ground almost any way that seems possible and fast. This is dangerous, and its prognosis doomful. This is when people try to go lower when they should be going higher; it's when they try to auger down through a too low and small hole, or when they try to land in a field that hasn't been looked over carefully, nor the trees, wires, and what have you around it.

What you must do when fright takes over and panic begins to spread is get up where all obstructions are well cleared or, if VFR, stay where it's safe, even in a circle, and then, in both cases, take time to think the situation through carefully, get settled down, and after that to act calmly and precisely.

While it isn't a part of the emotions of weather flying, it is related to being in trouble, and so we'll talk about asking for help. A pilot shouldn't be bashful or reluctant to say he's in trouble or getting into trouble. The radio is there to aid pilots. There's plenty of help around and a good clear yell will bring it running. Say what's wrong, where you are, what's needed most. Use the frequency currently being used or an emergency frequency. The calmer one can be while doing this, the easier it will be for someone to help.

THINK, FOR REAL

People rarely get into the situations we are talking about here, and weather flying isn't a desperate thing at all. But one reason why people do get into trouble is that they don't do enough logical thinking in advance. The preflight analysis of weather, the decision to go or not, the planning of fuel, and the cold decision of whether the airplane and pilot are adequate must be done logically and scientifically. Hunches don't have any place here—with the exception that negative ones cannot hurt, such as, "I've got a hunch that place will fold up, so I'm going to take more fuel." Actually that probably wasn't a hunch, but a judgment made after studying the weather.

But irrational optimism, like thinking, "It's going to be okay," on the basis of a keen desire to get where one wants to go, is taboo in this business.

If a pilot uses logical thought processes and keeps his emotions under control he will be able to handle tough problems. And if he trains himself to think the proper way, he probably will not get into bad situations in the first place.

Turbulence and Flying It

TURBULENCE COMES IN ALL SIZES from little choppiness to big, hard clouts. It affects us near the ground and up high where jets fly; even U2s at 70,000 feet find turbulence.

Unfortunately we cannot see the motions of air, and there is still doubt about how, exactly, the air moves when it's disturbed.

Flying gliders in mountain waves has made me wave-conscious. Flying over the sea, too, one cannot look down on the endless waves, whitecaps, and swells without thinking that the ocean below isn't much different from the one we fly in. They just have different densities.

A wave comes toward the shore and hits a line of rocks. The water sprays and tumbles over the rocks to fall down on the surface beyond them.

Is this much different from wind hitting the trees on the approach end of a runway, being thrown up and broken to descend in a confused manner toward the ground beyond the trees? No, I don't think it is.

And way up at 35,000 feet couldn't turbulence be a moving mass of air, in waves, coming up against denser air. The wave breaks and, if we could see it, might look like an ocean wave, the kind surfers ride.

All this comparison may be amusing to think about, but

how does it help? It helps in visualization. Our wave breaking over rocks, which we would have a wild time riding in a boat, can be much like the wind coming against obstructions; it will be disorganized, so that our airplane will roll, yaw, and pitch, and it will sink. We must be prepared to combat this.

Visualize water rushing down a streambed and then flowing over smooth downward sloping rocks. It follows the contour of the rockbed.

A runway perched up high, with the ground sloping away from the approach end, is the same thing. Coming in to land we fly over the sloping area, where there the wind flows downward, and we sink as we fly into this downward-flowing air. A marginally low airspeed, plus aiming to land short, may suddenly find us touching down before the runway, below it, with a good chance of leaving the landing gear back on the slope as we slide on our belly along the runway.

What we are saying is that a good pilot visualizes the air motion around an airport, a mountain, or any obstruction near which he is flying.

As we fly by an airport on downwind leg, we should have in mind the wind direction, velocity, and gustiness. Then, with crafty eye, we look at the obstructions and visualize how the air is moving over them and how it will affect our airplane. We should be prepared for choppy air, a downdraft, or sometimes an updraft that will make us balloon when we don't want to.

Mountains can cause up- and down-drafts quite a distance from an airport, and one should look at the terrain for many miles in all directions and try to visualize its influence.

I experienced a good example of this when taking off, with my son, from Honolulu in a Cessna 402 for Tarawa, 2400 miles away. It was a ferry flight and we were overloaded with

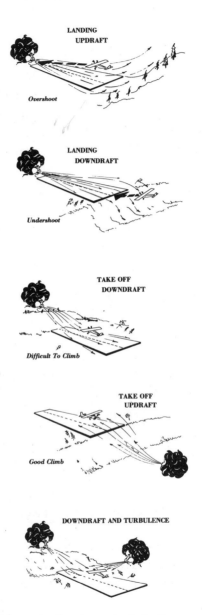

LANDING
UPDRAFT

Overshoot

LANDING
DOWNDRAFT

Undershoot

TAKE OFF
DOWNDRAFT

Difficult To Climb

TAKE OFF
UPDRAFT

Good Climb

DOWNDRAFT AND TURBULENCE

Some basic—but important!—views of wind effects around an airport.

extra fuel. Once in the air, after a long run, we made a right turn and headed out to sea. Our climb was very poor, but after flying a few miles away from the island it suddenly increased to the normal rate for the load we had. This was to be expected because the ground slopes up to about 2800 feet six miles northeast of the Honolulu airport. The pleasant northeast trades were flowing down the slope of this Koolau Range, as it's called, causing a large area of settling air we were trying to climb in. It kept us down until we flew out of it, eight to ten miles from the mountains.

KINDS OF TURBULENCE

Turbulence can be categorized according to its location. These are the places and areas pilots find turbulence: near the ground; in those obstructions we talked about; in the convective layer of the atmosphere or, in simpler words, in the haze below the inversion; in clouds, of course, and the more unstable the cloud the rougher; and finally in clear air. This last is not to be confused with the turbulence in the convective level; it is turbulence in clear air well above the convective level, generally in connection with jetstream. But there are other things that may cause it, and we'll go into them later.

HOW WE FLY TURBULENCE

We should talk about the ways we fly an airplane in turbulence. Basically it's a matter of not fighting, but rather letting the airplane have its way with little displacements it wants to make. We don't sit there and madly fight the stick or wheel.

This, of course, has its limits, and there comes a time when we say, "Whoa, baby, you've gone far enough." Then we move controls and make the airplane come back where we want it.

Since an airplane pitches, rolls, and yaws, we might look at turbulence from those aspects. Let's start with pitch—the up and down movement of the nose.

A hefty gust will make the airplane's nose change position. What we want to do is keep the nose where it should be, and that position is the place on the horizon, real or artificial, for the speed we are trimmed. We fly this attitude and keep the nose near that point on the horizon.

If we take off or put on power, this position will change, and we must retrim. But these trim changes are small. What we want to avoid is large displacements of the airplane. In clear air on a calm day try to see what positions the nose takes for different speeds and with different power settings. One quickly learns that big power changes make big attitude changes. This is a clue, of course, that juggling power during turbulence in large amounts will result in big attitude changes and general confusion in flying. We don't want that. It's best to know the speed one wants, trim for it, and then ride it out, keeping the nose very near this trimmed position.

On a clear day one should try, and then memorize, pitch settings on the Artificial Horizon vs. power settings and trim, for everything from best turbulence penetration speed to climb after takeoff. They will vary with load and altitude, but will be valuable as something to grab quickly when needed—at least as a starter.

Altitude will vary. If it's very rough and you are worried about the structure, let it vary! Otherwise make the smallest possible power adjustments and keep the flying technique as simple as possible.

Roll and yaw are mixed together to different degrees in different airplanes. Push a rudder pedal and the airplane yaws; as it yaws, it will begin to bank—sloppily, but it will. In a swept-wing jet this tendency is very strong. Push the rudder and immediately it starts to bank, too. Bank the airplane, on the other hand, without moving the rudder, and its nose will eventually start to yaw around.

Another thing that happens with bank is that the nose goes down. Bank the airplane, and the nose starts down; unbank it, and the nose rises. The important time, however, is when it goes down. That's how spiral dives begin if a pilot is on instruments in rough air and is not experienced. The airplane banks, and the nose wants to go down. If the pilot pulls back without doing anything about correcting the bank, the bank steepens, the nose drops more, and the speed goes up. It's a spiral dive off and running! It happens very fast. Many airplanes in this condition will, when improperly flown or not corrected, reach red-line airspeed in something like five seconds or less!

So, keep the wings level! It's simple, really, to keep the wings level and keep the nose where it should be on the horizon. Do it in a relaxed way with easy pushes and pulls, not jerks and shoves.

CONVECTIVE-LAYER TURBULENCE

Now let's talk about turbulence in the convective layer, the stuff below the haze level. Today's a day for it. A front passed yesterday and from my study window, across the rolling country, the visibility is good in a brisk northwest wind. There are small widely scattered cumulus drifting rather fast across the sky.

Just a glance at all this says it will be rough from takeoff until I get on top of the CU—probably only about 5000 to 6000 feet because it's November, and in colder weather the tops are low.

There are mountains in my country with farms, fields, and woods. As we take off and climb, it will be rough; turbulence will start pushing us sideways as soon as we leave the ground,

Fly above this kind of cloud. Beneath these cumulus it will be rough and uncomfortable, while on top the air will be clear, smooth, and cool. (NOAA PHOTO)

jolting and jarring. That will last until we reach 800 to 1000 feet because of the turbulence of the unstable air being joined by the turbulence of that strongish wind bouncing over the hills, trees, and fields. So there are two types of turbulence, orographic turbulence caused by terrain and instability turbulence within the air mass.

Above 1000 feet it will still be rough, quite rough, but some of the small, jiggling, jolting, fox terrier kind of action will be gone as we get up out of the orographic influence.

We reach the CU level, climb up above that, and suddenly, like flipping a switch, it is smooth. The haze is below us, and at our level of visibility is truly unlimited. It's the place to be.

But let's go back and wallow around in the rough air below. How rough it is depends on how unstable the air is and how strong the wind. The fresher the air mass (that is, the first day or two behind a front), the stronger the turbulence. The air is colder than the ground, which heats it and sends it upward in fast climbing thermals. These are the days glider pilots become giddy. The sky is full of lift, and they make records and go long distances.

But in an airplane each thermal we go through is a bump. Some of them are pretty darn strong bumps. Out west it's not uncommon to fly a glider in thermals going up more than 1200 feet per minute. Hit that column of rising air at airplane cruising speed, and it's a good jolt.

IT'S ROUGHER THAN YOU THINK

I've often thought, while bouncing along in the convective layer, that if I were on instruments, near a thunderstorm, hitting bumps this hard, no stronger, I'd be quite concerned, slowed down, and alert to use best turbulence-flying technique. But because it's clear and sunny, we don't think much about it. Sometimes we should.

One special time is during descent. We often see a pilot push the nose down and let the airspeed increase right up until the airspeed needle is tickling the red line. It's smooth, and he's making impressive time. But he slips down through the inversion into the convective layer and suddenly he is hitting really solid bumps. The combination of very high speed and the strong bumps is certainly putting a heavy load on the structure.

An alert pilot will reduce his speed before whamming into the roughness below the haze level and then, after he has felt the bumps and tested their intensity, decide what speed to use the rest of the way down.

There's little reason to stay down in the convective layer and bounce around if you can top it. In the East this is generally fairly easy, because the top of the layer isn't much over 7000 in winter and 10,000 in summer. In the Far West it's another matter. Because of the high ground and strong heating from the semiarid land, the haze level will often be above 15,000 feet. Then you need oxygen or pressurization.

Flying in the constantly bouncing air is fatiguing, and for passengers it's uncomfortable. The constant bouncing doesn't help the aircraft structure either. This isn't a serious point, but it is worth considering.

DUST DEVILS

On windy, rough days, especially in the west, dust devils form. These look like miniature tornadoes that work across the countryside kicking up dust in a swirling cloud. They are easy to avoid, and it's a good idea to do so. On landing one could get roughed up near the ground and settle drastically or hook a wing as one is tossed around. The action feels much like prop or jet wash; it has that uncontrollable feeling. Even after one is on the ground, the dust devil can lift a wing or flip an airplane over.

Big dust devils are obvious, but smaller ones may only look like a gossamer, funnel-shaped ghost skittering along. If we weren't looking for it carefully, we might fly into it.

When dust devils are forming, it's important to have airplanes on the ground well tied down. A dust devil can turn over a parked airplane easily.

TURBULENCE NEAR MOUNTAINS

Flying during strong wind conditions in mountainous areas, particularly when the air is unstable, can result in some very rough rides. The windward side of a mountain range will have predominantly rising air. It will not be consistent, but rather choppy and gusty. The gustiness will show itself in great ballooning climbs that will make you feel you'd like to reduce power, followed by sinking feelings as the up-gust diminishes. Often this variation will simply be the result of flying from one mountain contour to another. Mountain ranges are rarely ruler-straight. They have cuts, and the range twists so that its face is at different angles to the wind; the mountain's slope angle will differ from place to place too. All these irregularities create variations in the winds over the mountain. Glider pilots who have done a lot of ridge-soaring learn this, and an experienced one can look along a ridge and tell what kind of lift and turbulence he will find and where.

On the downwind side of a mountain the air will mostly be

going down. It will be chopped up and rough. The place to fly is well away from the mountainside.

Given a choice, I'd rather fly on the upwind side of a ridge than the downwind side.

We don't always fly along mountains, but often we cross them at right angles. When we do, it's important to note if we have a tailwind or headwind.

Say we are flying low, 500 feet or so above the ridge tops. With a tailwind, the airplane will climb as we approach each ridge. We are being "zoomed" by the air flowing up the slope. But when we pass the mountain ridge and fly to the downwind side, the airplane will sink, and we'd better be prepared for it.

Flying the other way, toward ridges into a headwind, is hazardous because approaching the ridge we enter its downwind side. We could also call this the downslope side. We will sink, often below the mountain top, and need power to keep from getting too low and too slow. We shouldn't just pull the nose up and let airspeed drop while trying to climb. Use lots of power and get up.

But it is best never to fly low near mountains, particularly if the wind is strong; we should be especially careful flying toward them into a headwind, since one can sink and get in a dangerous position.

A hazardous place to fly is between two ridges so closely spaced that downward-flowing air from one ridge is turning to climb up the next ridge. The air movements will be confused, and there will be lots of sink.

When flying along or across a mountain range, realize that the wind speed will increase where cuts and passes go through the range, because of venturi-like action, and the turbulence will become worse.

The wind speed over the top of a mountain increases in much the same way that air flow increases over the curve of a wing. In the extreme, Mt. Washington in New Hampshire has recorded winds over 200 mph! It wasn't the general wind going that fast, but rather the wind accelerated by the curve of the mountain top. Almost any mountain will have this effect to some degree. Where the wind speed increases there will be shear and turbulence on the top and the downwind side.

Of course, what all this means is that we shouldn't fly close to mountains when the wind is strong and the air unstable, especially the downwind side. If you are close to a ridge, never, never make a turn toward the mountain; always turn away. To really respect and learn about flying in mountains I strongly recommend ridge flying in a glider . . . it's fun, too.

MOUNTAIN WAVES

Mountains create another type of turbulence that extends above the convection level; it's connected with mountain waves, sometimes called lee waves or standing waves.

Fast-moving air hitting the side of a mountain causes waves in the air downstream of the mountain. They are like waves in a river. Water flows over a submerged rock and causes waves for quite a distance downstream of the rock. A mountain wave is the same thing—the airflow is the water, and the mountain the submerged rock. The waves in the sky sometimes extend to great heights. The frontside of the wave is going up, and glider pilots like to hunt for this area and make altitude flights. Paul Bikel broke the world altitude record in a glider by going to 46,267 feet in a wave off the Sierra Nevada Mountains of California! From 4000-foot mountains in Vermont we get as high as 29,000 feet. I've felt wave influence at 7500 feet over a 700-foot hill.

Let's visualize these waves. They form downwind from the

mountain, not very far off—in fact almost over the peak of the mountain. The waves repeat downwind with second, third, and fourth waves and often lots more. The waves go up on one side and down on the other. We often discover them by noticing the airplane wanting to go up. We push the nose down and get increased airspeed without adding power. We are in the up part of a wave. The air is generally glassy smooth.

On the other side we get a smooth downdraft. We have to pull the nose up and add power to hold altitude. This area is smooth too. So far so good.

But under the wave we find turbulence in an area called the rotor, where the wave air conflicts with the undisturbed normal air below and causes a tumbling, rolling, chopped-up, confused mess. It is rough; how rough depends on the strength of the wind, the mountain, and how much of a wave there is. It can be very rough! In severe cases this rotor has torn an airplane apart; it's not a thing to fool with.

How do we know it's there? First, with strong wind flow across a mountain, especially after a front has passed, we can expect wave action. Looking up into the sky, we may see lenticular clouds. These are a characteristic sign of waves. They are long, slim, lens-shaped clouds; but the real key is that they do not move as normal clouds do. The reason for this is that the rising side of the wave cools air to its condensation temperature and forms the cloud. As the air starts down the backside of the wave, the pressure increases, the air gets warmer, and the cloud disappears. So the lenticular cloud forms and disappears on the wave, staying with it, and doesn't drift along with the wind. There aren't always lenticulars with waves, but they are still worth looking for.

In the valleys behind the mountains the rotor often extends

Beautiful lenticulars that tell a wave is working in the sky. They look smooth and peaceful, but below it can be very turbulent. (NOAA PHOTO)

almost to the ground; the air near the ground is rough, and one can see clouds of dust swirling into the sky, if it's a dry dusty area.

I've flown through a rotor in Vermont that was almost on the airport at Warren. Descending to land, it was wildly rough.

Flying downwind of mountains, we want to be especially careful when the possibility of a wave exists. Look for it after a front and with wind across the mountain range. Sometimes we can get the combination of air-mass instability and wave-rotor action together, and it makes a mighty interesting ride!

Rotors can lie fairly high in the wave structure, but the most turbulent rotor will be found at the mountain level and below. If the mountain is 4000 feet high then the rotor will roughest from 4000 feet down. To be on the safe side we should add 2000 feet or so to the mountaintop altitude to assure being out of the worst part of the rotor. If you can see the rotor clouds, fly above them. In bigger mountains it would be advisable to add extra altitude above the mountain height, say 5000 feet more. But the best way is to avoid the area altogether.

Waves can be created by any size mountain, but they are not always visible by signs of a lenticular because there may not be enough moisture aloft to form a cloud.

The rotors, however, will sometimes show themselves as little pieces of shredded cumulus-like cloud. There may be moisture in the lower levels that will support, or almost, small CU, while there is not enough aloft to make a lenticular. So if you see CU on the downwind side of a mountain, generally not much higher than the mountain, and when you study them carefully you see a shredded, moving, turning-over appearance to the top side of the cloud, look out! That's probably a rotor. Sometimes there will only be a few small, half-formed, gossamer-like wisps of cumulus that you almost "feel" rather than see, but if we're watchful, and see them, they can tell an important story. Those innocent-looking wisps may signal a very rough rotor. Any atmosphere like that downwind of a mountain should be cause for suspicion and, if one must fly through, a time to prepare for flight in turbulence.

When getting in the uplift of a wave, the air is smooth and might fool a pilot flying an aircraft with low power and no oxygen. He could go up at a fierce rate to a high altitude where oxygen is needed. The way out of such lift is to turn downwind. But then he'd get in the *down* part of the wave, which is also generally smooth. The descent rates may be high, more than he could overcome with his engine. Pulling the nose up and pouring on power may not be enough to hold altitude—a clumsy pilot could get in a mushing, stalled condition. It would be best to keep up speed and fly fast downwind to get out of the wave condition. He might be going down fast while doing this, but the combination of airplane high speed and the tail-wind would get him out of the down current of air quickly. The pilot might go through a number of waves in his down-wind dash, but he'd be getting out of it the fastest way possible.

A special situation is a single mountain that sticks up alone, like Rainier in Washington State or Fuji in Japan. Under strong wind conditions the downwind side can have a com-bination of wave and vortex. The air spills over the top and around the sides and sets up a great whirlpool action down-wind that has severe turbulence. A 707 was torn apart in this condition near Mt. Fuji. So it's well to be very wary down-wind of peaks that stand alone enough to create this situation.

Mountain waves are interesting, and playing with them in a glider is not only fascinating, but a good education for the airplane pilot. I've landed many times at Milan, Italy, which is snuggled against the southern side of the Alps. Because of glider experience I'm wave-conscious and have studied and learned which conditions make the biggest waves in the area; generally a quick look at the winds aloft pattern tells me if I have to be extra careful while descending into Milan. In-cidentally, the strongest waves there come with southwest winds from the direction of the French Alps.

Wave action is something we can use to advantage on cross-country flights if we're paralleling mountain ranges.

On top, particularly after a cold front passage, we feel periods of descending air, and our airspeed drops way off as we try to maintain altitude. At other times we seem to get a great

Action on the downwind side of a mountain range. In the clear air immediately to the left of the mountains, there is a downdraft—a very strong one that only powerful airplanes could handle, and the pilots would have their hands full. Where the dust is rising, the air is moving up in strong wave action and is very turbulent. It would be best to avoid the entire area. (PHOTO BY ROBERT SYMONDS FROM DE VER COLSON COLLECTION)

An unusual and dramatic picture of a vortex situation downwind of a lone peak. If there weren't enough moisture to form a cloud, this condition would not be visible and one might fly into it unexpectedly. Lesson: stay away from downwind of lone peaks when winds are stronger than light. (PHOTO BY JOSEPH SCAYLEA)

boost and an increase in speed way over normal cruise. What is happening is that we're flying in and out of waves, perhaps only mild ones. If we watch the top of the clouds we'll see undulations like swells on an ocean. If we eyeball these and fly to stay on the upwind side of a "swell," we'll be in rising air and going fast. Moving our course around to fit these upflow areas may well be worth small detours.

At times we may feel this condition in clear air when we cannot "see" the waves. If so, and we're in descending air, a little exploring by moving our course more into the wind will find rising air. This awareness of the air's motion may save both fuel and time. It's the stuff glider pilots work with and it opens an exciting new concept and knowledge of the sky.

TURBULENCE UP HIGH

Sometimes turbulence that is not CAT is felt well above the convective layer, at 12,000 feet and up. This is generally light and associated with overrunning warm air preceding a warm front. It may also be there because of other air-mass changes. Often a look up will show altocumulus clouds.

THE BIG HIGH ALTITUDE TURBULENCE

Now let's talk about clear air turbulence (CAT). This type has been a bogeyman in high flying for some time; it's been popular in the press and is often called the "airman's enemy" by the dramatic types. Let's look it over calmly.

Most clear air turbulence is associated with the jetstream. The jetstream is a hose-like band of high-speed winds at high levels. It doesn't string out exactly east and west very often. Generally, it wanders in a serpentine fashion, so if one flies east or west, he may cross it a number of times.

The jetstream is up in the high latitudes in summer, 60 degrees or so, and down south as far as 20 degrees in winter. The hose-like part is the area of maximum winds that sometimes blow 200 knots and have reached 300 knots! There are also strong winds for hundreds of miles to the side of the special high-speed jet. The high-speed winds spread over a wider area to the right of the jetstream, looking downwind, than they do to the left.

Where these high speed winds rub against lower-speed winds, there is tumbling, turbulent air. Where air-mass densities change, as in a high-level front, it's rough. Where a jetstream first slams into slower-moving air ahead of it, it's turbulent too.

Meteorologists can locate the jetstream rather accurately, but they cannot pinpoint exactly where the turbulence will be: they can tell you only in general terms. After spending 8000 hours above 30,000 feet, I can assure you that sometimes it will be rough when they say it won't and smooth when they say it'll be rough.

How rough is it? Moderate at times, perhaps severe by some people's standards. Personally, I think the danger of the turbulence itself is overrated. It's uncomfortable and at times disturbing, but it isn't anything that can't be handled.

It is possible, however, to get into trouble because of control problems. Up very high the airplane has low thrust; it's mushing along at a higher than normal angle of attack. It's squirrelly, as the expression goes, because its Dutch Roll tendencies become more pronounced. In turbulent air, almost all of which is caused by shear, airplanes can approach stall and want to go off on a wing. It takes careful flying to keep level and under control.

This isn't a desperate situation if the pilot is on top of it and realizes that he may have to give up some altitude to keep control. This nervous area is why airlines try not to operate above any altitude where the airplane is close enough to stall, so that a 1.3 or 1.5G bump might cause the airplane to exceed the stall angle of attack.

Yaw dampers are a help in these conditions. They aid the pilot and are a control augmentation or, in a small way, a fly-by-wire concept in which electronic gadgetry makes flying easier. Modern military aircraft are loaded with it; so are the Concorde and space vehicles. There's nothing to be ashamed of, or feel inferior about, because a gadget can fly better than we can. It's a great thing and should be used with gratitude and enthusiasm. There will be much more of it in the future.

The formula is to fly lower and fly as one would fly any turbulence, by attitude, with wings level and pitch and power changes at a minimum. Doing all these things will make CAT a nuisance and not a disaster.

WHERE IS IT?

Despite the difficulty in forecasting CAT there are some cues the pilot can watch for to tell him where CAT may be.

Before takeoff he should have a look at a high-level pressure chart—the 300 millibar one, which is about 30,000 feet. Its wandering isobars will show where the jetstream is apt to be. It's like any pressure chart, with the distance between isobars telling how strong the wind is—the closer the stronger.

In studying the wandering curve of the isobars, the pilot should note especially where the isobars bend and change direction. A trough, for example, will have the isobars oriented from northwest to southeast on the west side; then they will turn a corner and orient themselves southwest to northeast farther east. In the area of the corner where the wind changes direction you can almost count on finding CAT.

Then check the isobar pattern. Let's say the isobars are all crowded together in an east–west direction; the wind is moving fast. But farther east the wind slackens, and the isobars fan out. It will be rough in the area where the wind velocity changes. If the velocity changes faster than 40 knots in 150 miles, it's a sign there will be considerable rough air.

Jetstreams above mountain wave areas have extra turbulence, and the type found in the rotor will be found at high altitude, but not with the viciousness the lower rotor has.

The Weather Bureau also publishes information telling wind velocity change with height. Generally, it is considered that if one sees a wind change of four knots or more per 1000 feet, it will be a rough area. This I find hard to count on. Sometimes it seems to work, but I've seen occasions where the four knots and everything else said it was going to be rough, only to fly through and never hit a ripple. If this figure coincides with other turbulence cues, like bends in the isobars, it seems more valid. At any rate we watch these areas with suspicion.

THE TROPOPAUSE AND CAT

Perhaps the most interesting thing to study in advance is the height of the tropopause, which tells us a lot.

The tropopause is the place where the troposphere, in which we live, ends and the stratosphere begins. The most noticeable characteristic difference between them is temperature. As we know, the temperature decreases with altitude in the troposphere and keeps right on doing so until we reach the tropopause, where the temperature stops getting colder. This, theoretically, is at about 35,000 feet in the average latitude of the USA. Above the tropopause the temperature in the stratosphere is considered to be about 56 degrees C below zero, but it varies considerably.

To make life interesting, the trop, as the trade calls it, wavers up and down with the passage of lows and highs under it and can actually range in altitude from the low 20,000s to the 40,000s right over the USA. Normally the trop is high over the equator, at about 55,000 feet, and low near the poles, at about 24,000 feet.

THE TROPOPAUSE IS IMPORTANT

A couple of things happen at the tropopause. The temperature goes up or holds steady, as we said, and there is a choppiness as you go through the tropopause into the stratosphere. These things do other things. The temperature change is an inversion just like one 1200 feet above the ground early in the morning. The inversion makes sort of a cap, and if there's a jetstream, it will be fastest right under the tropopause. If this band of strong wind contains turbulence, it will be roughest right at the trop and for a few thousand feet below it. The depth of the region where the turbulence and wind are strongest is a point of argument, but personally I like to give turbulence at the trop a 4000-foot berth—that is fly 4000 feet under it. That doesn't avoid all the turbulence, but it makes it less bothersome.

Above the trop, in the stratosphere, the turbulence dies out within a very short altitude range—1000 feet or less. If you get *up* out of the trop, it generally becomes smoother quickly. Wind decreases too; if one has a strong headwind it is often possible to see it drop off dramatically as soon as we climb into the stratosphere. If we're trying to avoid headwinds, and the temperature is low enough to operate up there, the place to fly is above the trop—and it'll probably be smooth, too.

Since jets fly between about 28,000 and 41,000 feet, it's simple to see that they wander in and out of the tropopause as they cross over high and low pressure areas. This means that our current jets fly at the most annoying altitudes. Everytime you pass through the trop it's bumpy, and if there's a lot of wind or turbulence, it may be very bumpy.

Nuisance number two in the trop is the temperature change.

A simplified view of a tropopause chart and visualization of its effects on a Seattle-to-Chicago flight at 39,000 feet.

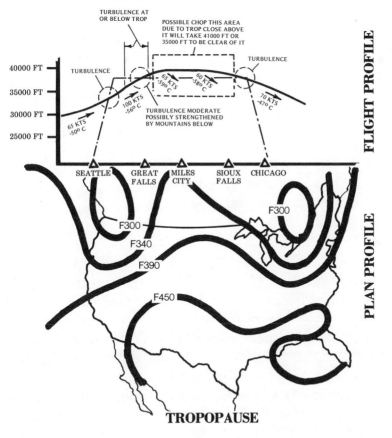

You take off for a long trip with a heavy airplane and plan to cruise at 33,000 feet. The weather charts say it's cool there, and the trop is well above you. But once in the air ATC says to climb to 35,000 feet. As you go through 34,000 feet, there's a little choppiness, and you can see a slight haze level. Your eyes dart to the outside air-temperature gauge just in time to see it go up to a "warm" 44 degrees C below zero. The airplane feels mushy in the "warm" air and life is no longer beautiful. You are above the trop.

This is an inconvenience. You may stick it out and stagger until the weight lowers through fuel use or, if it's too difficult, you call ATC and tell them you just cannot fly up there. Then they have the problem of working out something lower for you.

All this makes studying the tropopause charts very important before flight. We're looking for these things:

1. Trop heights
2. Trop temperatures
3. Trop wind direction and velocities
4. Vertical shear values

We are interested in the trop level especially in the area of takeoff and where we'll reach climb altitude. If the trop is supposed to be 33,000 feet and it's temperature −50°, we may not be able to climb higher immediately, even if we'd like to, because the temperature above the trop will be −50° or warmer. We may have to stay lower until we fly into a higher trop area, or until we burn off fuel, are lighter, and have better altitude capability.

Vertical shear will show as boxes on the trop chart with numbers in them. We might see a value such as 4. This means the wind velocity changes 4 knots each thousand feet. A value of 4 or more is generally considered enough shear to call for turbulence, but as we said, it doesn't always happen.

A trop chart has many characteristics similar to other pressure charts. If the trop levels change greatly over a small distance, the winds will be high speed and it'll be rough. In a steep slope like that, the temperatures will change quickly and, as in a lusty cold front, there will be rough air. If the trop levels make a sudden change in direction, it will be rough there too.

The trop chart shows how high we'll have to climb to be out of rough air and strong winds. If we can climb that high all this is very pleasant to know.

For high flying the trop chart is as important as the surface chart in telling weather. Near, in, or above the trop is where high-flying aircraft spend their time, and familiarity with the phenomena of the trop is of utmost importance.

If we don't get to a place where we can see trop charts, we should ask the briefer we talk to via telephone, or other means, questions that will tell what the trop charts show.

SHEAR

Shear is a wind effect and the cause of almost all turbulence.

We feel it when flying from one strength wind to another. In the case of flying from a strong headwind into a lighter wind, the airplane loses airspeed and must get it back. At cruising speed this isn't much of a problem; but at low speeds during an approach to landing it can be significant.

If we are approaching at a slow speed, we are not too far from stall. A sudden airspeed loss, caused by flying out of a headwind, brings us suddenly much closer to the stall. We sink. If we don't do anything about this, it will take a long time to get the speed back—many seconds, even a half-minute or more. In that time we might get too low and hit short of the runway.

When this speed loss occurs, we have to get on it right away. We pull the nose up to keep from sinking and put on more power to overcome the drag from the pullup.

The pullup can be quite severe in an extreme condition, and it may be necessary to pull up almost to the stall angle of attack to get up and away from the ground. If the down draft is strong, the airplane's angle of attack will be low and it will be necessary to change that to a big angle in order to get the necessary lift. The pitch angle—what you see on the horizon, be it true or artificial—will be quite large and it may look pretty desperate, but the actual angle of attack—the relative wind—will not be that extreme. During all this, of course, maximum power must be used, throttles right up to the limit!

The task is not to exceed the stall angle of attack, although close to it, and one will have to rely on his sense of feel and the stall warning device on the airplane. We are describing a very severe condition such as that which will occur in the gusts and downdrafts from nearby thunderstorms. It would be much safer and easier if we didn't try landing with thunderstorms close to the airport!

If strong shear is encountered and large attitude power changes are used to combat it, a serious hazard then arises as the shear stabilizes and the pilot attempts to continue the landing. The airplane is trimmed for a steep pitch attitude and excess power. To re-trim, get the power off the right amount, and land is very difficult flying, especially in airplanes that require large stabilizer travel. Shear also often has a crosswind component which will quickly push the airplane away from the runway center line. So the pilot not only has to do some difficult trim, power, speed, and rate-of-descent juggling, but he has to turn the airplane and get it lined up with the runway, generally at low altitude. All this is very tough flying and probably where shear accidents are actually caused. So if one is fortunate enough to recover from the shear he shouldn't press his luck by trying to get lined up and land. The smart maneuver is a go-around!

Shear is easier to cope with in a propeller airplane than in a jet. The propeller airplane, when we pour on power, gives thrust; it also produces an immediate increase in airflow over the wing from the propeller wash and therefore adds lift quickly. The jet engine provides thrust only, and it takes longer to regain speed. While all pilots have to be alert for shear, the jet pilot has to be particularly so.

WHERE IS SHEAR?

Shear happens in various ways. Wind velocities change as we descend (winds are usually stronger aloft and weaker near the ground), producing shear. Shear also occurs because of shifts in wind direction. In thunderstorms, when a front is approaching an airport, we can easily get a wind change from southerly to northerly in an instant that will make landing an exciting adventure.

More subtle, however, is a wind shift as we descend that changes a headwind into a crosswind and effectively makes the airplane's velocity lower.

A few hundred feet above the ground the wind follows the isobars. That's gradient wind. But the surface friction of the wind running over the ground causes the wind to flow more toward the low pressure; so the wind near the ground, instead of being easterly as the gradient wind might be, will swing toward the northeast and north as we descend. This, of course, cuts down its effective velocity and so produces a shear effect with an airspeed loss.

It's a good idea, when landing in a crosswind, to be conscious of what the gradient winds are. Is the wind aloft more of a tailwind in the direction we're landing, or a headwind?

If it's a tailwind, we may overshoot or land too long. For example, we are landing to the west. There's a north wind on the ground. The wind aloft is east at about 40 knots at 1000 feet. We are making our approach with a whopping tailwind. Our ground speed is high. As we descend, the tailwind decreases, but we only have a short period in which to get rid of a lot of groundspeed before we get to the runway. We tend to overshoot, and so we push the nose over to get down to the runway's end. Our speed increases, and lift becomes excessive. It's difficult to touch down on the runway; it would be wise to abandon the approach.

So we pull up and try landing to the east. Now we have a headwind aloft, but as we descend we lose the headwind and some airspeed. Now we'll have to correct for sink and overcome the tendency to get below the glide path.

It's obviously important to consider what the gradient wind is as we let down to land. It will give us an idea of what to expect during the approach. Will we be scampering and diving to get down and land within the airport—a tailwind condition? Or will we be pouring on power and lifting ourselves up to keep from undershooting—a headwind condition—when the wind decreases as we descend?

Sometimes the difference between the gradient wind and the surface wind can be quite high, even more so at night than in daytime. A nighttime inversion makes things seem still on the surface, but just above the inversion the wind can be swishing along at 50 knots if there is a strong pressure gradient.

We must therefore be wary of large differences between surface winds and winds aloft when the gradient is tight and there is a strong inversion.

Warm fronts can create strong shear conditions when things on the ground seem tranquil. First, let's review the wind directions on both sides of a warm front as well as above and below the frontal surface which slopes, from the front, generally toward the north, about 300 to 1—which says that 300 miles north of the surface front the front aloft will be at 5280 feet above the ground. The winds, north of the front and below the frontal surface, are north and easterly, often rather light. South of the front, and above the frontal surface, they are south to southwesterly and often quite strong.

Now let's place a warm front 30 miles south of Kansas City, where we will land. The surface wind at the airport is a gentle 10 kts or less and northerly, but above 600 feet the wind is southwest and 50 kts! So as we descend toward 600 feet we have a tailwind of 50 kts. Our rate of descent will be high and so will our groundspeed. We're all set for an overshoot.

A warm front is not especially bad because we may have to go around, but it becomes serious when the pilot doesn't go around and tries to make a landing, touches way up the runway and goes off the end into who knows what!

Overshoots also may put us in a position to undershoot! Descending fast, trying to get down to the glide slope, we pull off all power. Then we run out of that strong tailwind. Our groundspeed slows, but airspeed increases and we go above glide slope. In an effort to get back down our power is still reduced. Then, sooner than we realize, the airplane goes below glide slope and we are undershooting! Now it's pull the nose up and pour on power. Quite a juggling act, especially on instruments. The engine, if it's a jet, may be spooled down and take a long time to come back up to full power after the throttle is opened. This is when we risk landing short, taking out approach lights and maybe more.

So winds only a short distance above the earth can be very different from winds on the surface and tailwind-to-headwind, wind reducing in descent, wind increasing in descent, or headwind-to-tailwind, which all require different recovery techniques, aren't restricted to warm fronts, but can be experienced with cold fronts, nearby thunderstorms, mountain downdrafts, and anything that can give the wind different vectors and velocities.

All this theoretical talk isn't much good if we don't know where fronts are or if we suddenly run into shear conditions and don't know why they are there. Then what?

First let's recall the two basic things about shear: A headwind on landing tends to make us land short when wind velocity decreases as we descend. A tailwind tends to make us overshoot. Those are the basics.

So what we want to know is how strong the conditions are and how much the wind will change during descent. In fancier language, what is the wind gradient? This is part of our weather briefing—to give thought to wind changes with altitude near the ground at takeoff and landing airports.

Once in the air, the sophisticated way to know wind gradient is by having an Inertial Navigation System (INS) which tells groundspeed. If the groundspeed is much lower than airspeed, a headwind, and the surface wind is not very strong, there's a steep wind gradient and we'd better be ready for that undershoot situation when we fly out of the headwind down lower. A high groundspeed means a tailwind and an overshoot. An INS can give actual wind direction and velocity so the pilot can compare it with the latest reported surface wind and know how much change there will be in descent.

All that's fine, but most airplanes don't have an INS, so what about the rest of us? If we're on instruments, coming down on ILS, it's fairly easy to get an idea of what's going on by our rate of descent.

Let's say our approach speed is 120 kts. To make good a 3° glide slope, without wind, we'll descend about 600 feet per minute. So if we find it only requires 400 fpm to stay on the glide slope, we've got about a 40-kt headwind. If the surface wind is 10 kts, then somewhere during descent we're going to lose 30 kts and have to get power on and scamble to keep from getting too low.

If our necessary descent rate to stay on glide is near 800 fpm, then we've got a 40-kt tailwind and it's going to be a scramble not to overshoot.

Required descent rates vs. groundspeed often are listed on the bottom of approach plates, but if they aren't, there's a formula to help decide what the groundspeed should be. It is: half the groundspeed times 10 equals the descent rate, or nearly so. Take our 120 kts; half that is 60, times 10 equals 600 feet per minute. This is 36 fpm short, but is close enough. This, of course, is for a normal 3° glide slope.

We can do it backwards: if we're coming down the glide slope at, say, 900 fpm, we can divide that by 10, which equals 90, and multiply that by 2, which comes out 180, and that's our groundspeed. If our normal speed is 120 kts, we've got a 60-kt tail wind!

These numbers are for true airspeed, which, if we're near sea level and the temperatures aren't wildly off standard, will be pretty good. If we're landing at a 5000-foot-above-sea-level airport, however, we'll have to compute what the TAS is for our 120-kt IAS approach speed.

The formula, put up somewhere handy in the cockpit, will also be useful to help pre-guess, when you know what descent winds will be, the descent rate you'll have to use coming down the ILS. For one's particular airplane, with which we always use a certain approach speed, it's possible to mark the appropriate groundspeeds opposite the appropriate descent rates next to vertical speed indicator.

A visual approach is more difficult because there isn't any exact glide slope guidance unless there's a Visual Approach Slope Indicator (VASI) which you can pick up well out from the runway. Staying on a VASI glide path, and seeing what rate of descent is needed to do it, is the same as coming down an ILS glide path of 3°, if it's a 3° VASI—and most of them are.

But an all visual approach, without a reference to help, means getting back to basics. If we can pick a spot on the ground we want to land on, and then watch it in relation to our descent path, we know how we are doing. If the spot climbs in the windshield we're undershooting, and if it descends we're overshooting. So it's a sort of visual glide path. If we notice it's taking lots of power and a higher than normal nose attitude, we've got a headwind. If the nose is down, however, and we keep wanting to pull off power, we have a tailwind.

It isn't always possible to pick a spot in poor visibility or way out from the airport. If it isn't, then it's eyeballing the situation and using one's good seat-of-the-pants feeling and visual cues that come naturally. If we're dragging in there's a headwind, and vice versa.

Another type of shear is caused by gusty winds. It's obvious that a wind blowing ten knots and gusting to 20 can give one a very quick airspeed change of ten knots. That's why, on gusty days, it is wise to carry extra airspeed to take care of these sinkings of the airplane on approach as a gust dies down and leaves one low on airspeed. Most people take their normal approach speed and add half the gust velocity as a cushion during gusty conditions, plus another five knots if the wind is above 15 or 20 knots.

Shear is often overlooked as a takeoff hazard, but it's as important then as in landing. There have been cases where a takeoff was made into a headwind with a strong opposite wind—tailwind—whistling along just above the ground, resulting in an accident.

If there is a front of any sort near the departure point, one should discuss the possibility of a sudden wind shift with the meteorological briefer, and if a briefer isn't handy, but a front or thunderstorm is, we should be prepared for a wind direction change after takeoff. A thunderstorm, realistically, is the weather most likely to be violent enough to make the dramatic wind shift that causes trouble—we say it again: a thunderstorm near the airport, landing or taking off, is a very real danger, and sometimes one that cannot be flown successfully! But don't underrate nearby warm or cold fronts.

What do we do about all this? If the front, or thunderstorm, is close and strong, let's wait until it passes. If we must go—which is difficult to consider—or we get caught because we didn't think it was that bad, then consider that a loss of 50 kts, or a severe downdraft, is going to require an excess of speed as soon as possible after getting airborne. We want to get that excess quickly, remembering, of course, not to fly back into the ground or get too low on instruments or at night, in the process.

Sudden airspeed changes due to shear, bumps, or whatever, affect the aircraft's pitch attitudes. Two points: When getting a positive airspeed increase, or up gust, the nose will pitch up; when going from a headwind to a tailwind, or down gust, the nose will pitch down.

So we're taking off into a headwind and suddenly, right off the ground, we get an airspeed loss, or down gust. Airspeed drops and so does the nose. We should be prepared for this and, as in all flying, hang on to the pitch attitude we want.

THERMALS

There's another effect which isn't strictly shear, but acts in much the same way. It comes from thermals, the kind glider pilots look for: rising pieces of air.

On a summer day we approach a runway; perhaps there isn't much wind at all, but before reaching the runway we get a bit of sink and have to pull the nose up and add power. Then we cross the runway and suddenly seem to have excess lift and float for a long distance before the airplane will touch down. What happened?

Chances are this was an approach to a paved runway surrounded by grass or other vegetation. If we could see air circulation, we'd notice lifting air coming up off the paved runway and then (because when some air goes up, some other air must come down to replace it) we'd probably see air sinking over the grass adjacent to the runway. On our approach we'd fly through the sinking air over the vegetation; then, over the runway, we'd fly into the lifting air that would want to keep us up. The opposite will occur at night, with lift over the warmth-retaining vegetation and sink over the cooling runway. That's why we sometimes go clump in a hard landing at night and wonder why. We flew into sinking air. The night effect, however, isn't as strong as the day.

The important part of all this is that when sink occurs short of a runway, be it due to speed loss from shear or a thermal, we must be ready to use power and keep our airplane flying on the approach path which we have selected, either visual or instrument, and not allow it to land short.

VFR—Flying Weather Visually

SOME OF US fly VFR because we don't know how to fly instruments. There isn't any doubt that a pilot should learn instrument flying. It makes his flying safer, more useful, and more enjoyable.

Knowing instrument flying makes for safer flying because a pilot doesn't get trapped in bad situations. If he's cornered, he can pour on power, pull up to a safe height, and think out what he wants to do. He doesn't have to poke along between mountains or duck TV towers when ceilings and visibilities are low. Most of the time he can be on top, in sunshine, enjoying his flight rather than nervously and dangerously picking along in reduced visibility.

In the early days of the airmail—the days of open cockpits, helmet, and goggles—the pilots didn't fly instruments; they dangerously stayed "contact" and strained eyes and nerves trying to see what was coming up. That was when the Allegheny Mountains were called Hell's Stretch. Pilots plowed into them regularly, leaving broken bodies and airplanes on the rugged slopes. In the first year of the airmail service as many pilots were lost as the service had.

It pays to be instrument-qualified, and every day it becomes a more necessary part of flying unless one is willing to limit the use of his airplane.

VFR

But there's still a lot of VFR flying; most flying is VFR really, and there isn't anything wrong with it if the limitations are known and followed. First, always, is the fact that in flying VFR one must SEE! Low clouds and poor visibilities make this difficult. Rough terrain makes it worse. A safety key to VFR is not to get in a position where the visibility is so bad you cannot see enough ahead to make a normal 180-degree turn and get out. It's worth trying on a clear day, noting points on the ground, to see how much room it takes to turn, then double that distance for a cushion while one reacts, in actual weather, to the idea that the visibility is too low and he'd better turn around.

In VFR flying it is especially important to be flying toward better weather. VFR flying demands a good knowledge of the weather situation, and a VFR pilot should study weather in advance as much as an IFR pilot. A nice day at the takeoff point doesn't mean it will be nice for the entire route.

The VFR pilot must be especially aware of any possible weather deterioration. If it goes sour for him he must stay "contact," and this can mean getting too low in reduced visibility—a perfect formula for trouble! The IFR pilot is interested in trends, but deterioration to him may mean simply shooting a lower approach or going to an alternate. He does not get into trouble as deeply or quickly as the VFR pilot when the weather goes sour. So the VFR pilot needs a good look at the synoptic situation: where fronts are and their expected movement, plus the en route and terminal forecasts.

Basic weather rules for a VFR flight are these: fly toward improving weather; do not fly toward approaching fronts;

Flying into the picture: IFR yes, VFR no. It's obvious the weather ahead is worsening. This was near Tulsa, Oklahoma, with a summertime warm front and thunderstorms. A pilot should have a good picture of the weather and actual reports before poking around in this.

especially do not attempt to "beat" a front to your destination. Be extra conscious of destination forecasts when flying toward evening, shorelines, heavy industrial areas, and mountainous regions. Also take special care in spring and fall when the days are warm and the nights cool, particularly if there's a lot of moisture on the ground and in the air near the ground, which we get when lots of rain has fallen, and particularly when there is snow on the ground which has been rained on or subjected to melting during the day. We all know the feel of those clammy, cold nights—they make fog easily.

If there isn't any frontal condition, but cloud decks are forecast because of postfrontal air masses, be alert to the fact that clouds in mountainous areas will tend to be overcast and close to or on the mountaintops. If the air mass is vigorous, there will be showers and low visibilities.

Flying in mountainous regions requires more visibility than in flat terrain because clouds tend to hang on mountains and blend in, so that mountains and clouds often look alike. The slopes are difficult to see, and one comes up on them fast, even at 100 miles an hour.

A good rule is that if you cannot see the mountaintops, with space between them and the cloud bases, it's a poor time to be flying. There are, however, some conditions where there is excellent visibility, 20 miles or more, and we can fly down valleys with the mountaintops in cloud. But the forecast must be solid gold to stay good! The first snowflake or raindrop and it's time to find a place to land. And always be certain there's an entry and exit to the valley, because you cannot climb over the mountains to get out. Flying into a cul-de-sac—a dead-end valley—is bad, bad news!

A POINT TO REMEMBER

A VFR point worth thinking about is that the closer one gets to the cloud base the worse the visibility. As the ceiling squeezes a pilot lower and lower, he tries to stay as high as possible, and in doing so he crowds the base of the overcast. But the base of a cloud isn't clean-cut; mist and shredded hunks of vapor, like a gossamer veil, hang down below the solid part. A pilot can be flying in this area and see the ground straight down, but ahead, because he's looking through a lot of this ragged stuff, the visibility is low or nil. If he dropped down a little, sometimes only 50 feet, the visibility would suddenly improve. If it doesn't improve in 50 to 100 feet, then the bottom of the cloud reducing visibility isn't the problem, and it will not pay to go lower; it might actually be dangerous because of getting too close to the ground.

SNOW IS DIFFERENT

This idea of going a little bit lower to improve the visibility doesn't hold true in snow. Often, in snow, the ceiling is quite high. Snow generally forms directly from water vapor without any cloud process, and one doesn't get "in" a cloud. In professional jargon this process is called sublimation. Because there isn't cloud in this kind of snow, there generally isn't any ice either. It takes the supercooled water in a cloud to make ice; no cloud, no ice (except in the case of freezing rain). Therefore, if a pilot is in snow with the ground visible, he can pretty well relax as far as ice on wings and propeller is concerned.

Snow can come from nimbostratus, and snow showers from cumulus. These clouds are easy to avoid—generally by flying on top. Nimbostratus, or stratocumulus, are the kind we find over mountains after a cold front. Snow showers in cumulus would be in unstable air, so on top, again, is the place to be, but tops may be high. The clouds that have snow must contain supercooled water, and they get that by the lifting caused by strong instability or in air that's been lifted mechanically by wind pushing it up a mountainside. But generally in large snow areas there isn't ice. If ice forms, you are either in a lower cloud deck, which you can top, or you are near the frontal surface. More of that later.

The heat of the engine and windshield may cause snow to melt and refreeze as slush, just as on an automobile windshield on a snowy day. Because of this a pilot has to keep his eye out for carburetor ice all the time and for ducts getting plugged up after a long period of flying in snow.

When flying in snow caused by overrunning air ahead of a warm front, one can see the ground straight down from quite a few thousand feet. Of course the visibility ahead will be very poor because we're looking through a lot of snow. Because the snow forms aloft and falls to the ground, the restriction in visibility that it causes takes in all the sky; going lower will consequently not improve visibility. People are sometimes fooled by snow and fly lower and lower because they can see the ground straight down and feel that they should be able to get "under" the stuff. In fact, they cannot get "under" it and sometimes are hurt trying to. A pure VFR pilot had better stay out of snow areas, because the visibility will be poor regardless of the ceiling. Once snow starts—even light snow—the visibility will become poor in a short time.

If a VFR pilot is prowling along under leaden skies, he will often have good visibility, but when snow starts and he sees the first white streaks go by, it's time, right then, to think about going elsewhere or finding a place to land. The first flakes of snow are almost always a loud and clear signal that things are going to get worse.

Sometimes, ahead of warm fronts, snow begins to fall very quickly in a fairly wide area, 100 miles wide or more, so that one must consider the unpleasant possibility that a 180-degree turn will not always take care of everything. All this relates back to the point that we should study weather in advance and know, in our mind, what could happen and which way to duck if it does. When precipitation begins, especially snow, a pilot shouldn't sit fat, dumb, and happy until things get bad; he should do something before that—either try to turn around or else go somewhere and land. Which is a good argument for position keeping: knowing where we are and where the nearest airport is.

KEEP CALM

When we say go somewhere and land, we don't mean to do so in a panicky fashion. The worst thing we can do, in any situation, is get frantic, dash for Mother Earth, and land in the first field that comes along. We ought to stick to airports, and ones that fit the airplane.

MORE SNOW

Snow, just to confuse things, can also be found falling out of clouds. This condition is generally found in air-mass weather, such as a stratocumulus deck behind a cold front. Snow falls from the clouds, and the clouds have ice in them. In mountains the higher ridges are in the clouds, and these are conditions a VFR pilot should avoid.

The mountains of Pennsylvania, after a cold front passage in winter, are generally cloud-covered and getting lots of snow, and the visibility is poor to zero. It's not a frontal condition, but nevertheless it's a tough one for VFR.

Low stratus clouds, giving poor ceilings, increase as one gets closer to a front. In these clouds there is snow and ice; the snow, however, will be falling through the clouds from some-place above them. A pilot will be in snow, but this time he will not see the ground except occasionally through breaks. There will be ice when he goes through the stratus clouds. At best this isn't a good place for VFR flying. We'll talk more about it later when discussing instrument flying in weather. We can say, in passing, that a pilot can pull up and get on top of these ice-producing clouds; he'll be between layers but mostly on instruments, because the snow will cut visibility practically to zero. It will be a white-gray world without reference—for instrument pilots only.

TV TOWERS

VFR pilots sneaking along in flat country have to be wary of TV towers that stick up dangerously high and are difficult to see in conditions of reduced visibility. The TV people also have a way of putting them on mountaintops, and it can be quite a shock to sneak over a mountain with clouds crowding down on it, only to be suddenly confronted by a massive TV tower and its supporting wires.

MAPS ARE IMPORTANT

TV towers are marked on maps and can be spotted if the VFR pilot knows where he is—which brings out the important point that VFR pilots must be sharp at navigating by map. Navigation is most difficult down low, because you don't see as far and there aren't as many clues available to help interpret a map; at low altitude the area your vision takes in is smaller even with good visibility, and, of course, in bad weather the visual range is even worse.

Sneaking along VFR requires moment-by-moment knowledge of where one is; a pilot who doesn't always know where he is should raise his ceiling and visibility requirements to a higher value—say 3000 feet and ten miles as a starter.

Constantly knowing position also means knowing where the nearest airport is to run for if things get tough. It may be one recently passed, off to the side, or not far ahead, but we ought to know where it is.

There's a great tendency for people to tune omnis or ADFs and fly along on the signals, knowing they are headed for the right place but never really knowing where they are! A radio facility chart doesn't take the place of a map! Even the sophisticated twin-engine pilot, when one of his engines gulps and shudders, may suddenly wish he knew where that nearest airport was. Maps, too, tell us important information about terrain heights, mountains and valleys, and a host of other things that a VFR pilot should know in detail.

To have useful terrain height information a pilot must keep his altimeter set to the barometric pressure of the nearest airport so the altimeter will tell with accuracy where he is in relation to the terrain.

If things get really tough and we are prowling down low, perhaps following a railroad, we have to keep in mind high-tension wires that may cross the valley that we and the railroad are following. Down that low, heaven forbid, we also find radio masts that aren't on maps and other man-made things that stick up in our sky. Crowding a mountainside too close could possibly cause one to clip the wires of a ski lift that may be difficult to see, summer and winter. A number of glider accidents have occurred in the Alps in exactly this manner. This kind of flying is desperate, not to mention illegal, and no one should be doing it; we mention it only in case one gets himself in such a fix and, desperate or illegal, wants out in one piece.

WHERE IS THE WIND?

Flying VFR a pilot must be moment-by-moment conscious of the wind, where it's coming from, and how fast. In mountains it's helpful to realize that there will be more cloud on the upwind side of a ridge—and it doesn't have to be a steep ridge. Even a gently sloping hill will give enough lift to the air to cause condensation and clouds on the upwind side in a high moisture condition.

On the downwind side there will be downdrafts, and if the wind is strong, they'll also be strong, and turbulent. One needs to be flying in clear air, with visibility, when he wrestles downdrafts. The airplane's altitude should be above that of the range, if clouds will allow. The worst downdraft spillage on the downwind side of a ridge will occur at or below the level of the mountains, perhaps a little higher than the ridge, but not much.

There are areas where winds converge. This can cause turbulence and, sometimes, enough lifting to make clouds. If the wind is flowing down a valley and there's a ridge in the valley, the wind will split as it goes around this ridge. On the downwind side the wind comes back together again, pushing against itself, creating an area of convergence and vortex.

We have to consider that cuts in mountains, ridges oriented in different directions, peaks, and valleys, all cause moving air to tumble and eddy in very irregular ways that are difficult to visualize. Wind coming down one valley and flowing into another will give turbulence and convergence. Caution is called for in mountainous areas, especially in areas where the mountains are arranged in hodgepodge fashion and not neatly lined up.

Valley winds tend to flow up or down the valley, so when considering general circulation don't think of the wind in valleys as representing the general wind direction and velocity —or don't plan landings in valleys based on what the general wind is supposed to be—take a positive look and know what the local wind is before takeoff or landing. Also, when climbing above the valley, we'll fly out of the local wind into the general wind structure, and that may give us some degree of shear. Since it will come near a mountaintop we may not have much terrain clearance and will want to be cautious. The wind direction aloft may help one's decision on which direction to take off in light surface winds. The direction should be so a turn will be into the wind aloft.

NEAR CITIES

If one is approaching an industrial area, town or city, a detour around on the upwind side will give better visibility. This rule will often apply to bodies of water, too, where a light wind can give more fog and lower visibility on the downwind side.

SUMMERTIME

While summer is a nice time to fly, it sometimes brings serious visibility problems to the VFR pilot. The sky may be cloudless, but the summer haze makes a cross-country by map reading and looking a tough chore. This is an especially good time to be on the upwind side of cities, because the polluted air mixes with the haze and makes the visibility much worse.

An inversion causes all this, where the temperature aloft gets suddenly warmer. The hazy air, stirred up by thermal

The effect of air pollution on visibility. This smoke extends many miles downwind, while upwind it is clear—showing why it is obviously best to fly upwind of cities and industrial areas. (NOAA PHOTO)

activity, can rise no higher than this altitude where the temperature increases.

Climbing, we come out of the mass of glop and find ourselves in clean, clear air with miles and miles of visibility. Looking down, however, we see almost nothing, and navigating is difficult. An occasional river glints through the smaze, the white ribbon of a highway, a piece of a mountain. It's all difficult to put together in the jigsaw puzzle of map reading. If we are omni- or ADF-equipped, we can use the radio for navigation and by looking at the occasional clues below know where we are, where that nearest airport is, and how we are doing. Without radio help it's a job of superconcentration, and often it's worthwhile to swallow pride and fly a slightly longer course to follow a visible highway.

THUNDERSTORMS AND VFR

Summer brings thunderstorms, of course, and the reduced visibility in haze makes them difficult to see. On top of the haze level the big cauliflower clouds can be seen clearly and be avoided. Down low, in the smaze, one doesn't know if the bad visibility is smaze or darkness from a closeby thunderstorm. Flying high, above the haze, ducking thunderstorms, a pilot needs to watch below too, because clouds may sneak in under him if he's close to a storm.

There's lots of information available about thunderstorm locations, intensities, and movement, and one shouldn't be bashful about getting on the radio and asking Flight Watch or an FSS what's around. Don't count on these reports 100 percent, however, because they may be old when you get them. But they are useful as a judgment factor.

Thunderstorms, at best, are things to keep away from. This is especially true for the VFR pilot.

A good look at the weather before flight will show if thunderstorms are going to be of the air-mass types or ones caused by frontal activity. Air-mass storms are scattered enough to tour around, staying in the clear. If they are frontal storms, the VFR pilot's place is on the ground!

Ducking around thunderstorms means staying out of them, and it means, too, giving them a wide berth. Turbulence outside a storm can be as rough as that inside. There's also danger of hail falling from the high overhanging anvil portion of the cloud into clear air. A general rule is not to fly over lower clouds or under higher clouds in the immediate vicinity of thunderstorms.

While weaving around, it's best to keep working into the wind simply to make any return to course shorter and your groundspeed faster in passing the storm. Compass headings should be noted: knowing the time on various headings is a help in figuring how far off course you may be going and which way.

An important point about headings is that if one seems forced to keep working in one direction, say southwest, and is unable to work back northwest because of storms, it's obvious that one is making a big end run. This means that the storms are pretty well lined up and there isn't any clear path on the desired course. It may be a front, a prefrontal line squall developing, or a bunch of air-mass thunderstorms that have lined up in a pseudofront. It's time to land or turn around and go back.

A pilot must consider that the storms he's going around may be right over the airport he's headed for. If so, he's got to be prepared to go elsewhere or else fly around out in the clear until the storm drifts away. This latter means he must have enough fuel and daylight.

Flying VFR at night around thunderstorms isn't a recommended procedure; the difficulty of seeing where clouds are is great, and in early evening at least the storms are probably worse and clouds more extensive.

A point to remember about air-mass thunderstorms is that they occur mostly in the afternoon. If a pilot gets an early morning start, he can get where he wants to go before the afternoon storms have built up enough to bother him.

There is much more to be said later about thunderstorms, but right here, while talking about VFR, these points are important, and the most important of them is to stay in the clear and always have a wide avenue handy for hasty retreat.

VFR ON TOP

Flying VFR doesn't mean one has to stay under clouds. If there are scattered clouds with a top that's not too high, it's better to be on top in clear air than down in smazy air worrying along in reduced visibility. It's also more pleasant because the air is smooth and cool.

You must be certain, however, that those scattered clouds stay scattered and don't become broken or overcast. A non-instrument-qualified pilot must be able to come down without going through clouds.

A good study of weather before takeoff can assure the pilot he's not flying toward frontal conditions or into mountainous areas where clouds tend to be broken and, at times, overcast instead of scattered. But if the clouds are scattered, there's no sense in sitting down low working and worrying unnecessarily.

USE THE RADIO VFR

Most airplanes are radio-equipped, even though they may have only simple equipment of doubtful strength. The VFR pilot can use this equipment, however, as an aid to navigation, for getting weather reports and for following, if he's filed, a VFR flight plan. This is good training for the future when one works to get his instrument rating.

There are two basic parts to instrument capability. One is knowing how to fly by instruments, to keep the airplane under control without outside visual reference. The second part is navigation by radio, handling radio communications for ATC purposes and obtaining weather information.

The pilot who cannot fly instruments can nevertheless learn and develop proficiency in navigation, ATC, communications, and weather gathering as he flies VFR.

He listens to scheduled weather broadcasts, learns when they are broadcast, where from, and on what frequencies. He can develop good habit patterns and useful knowledge.

He can navigate using radio, following a track, crossing checkpoints, checking speed—remembering, of course, to keep tabs with a map, too.

He can file VFR flight plans and keep in touch with Flight Service Stations en route, obtaining the altimeter setting of the airport nearest him so his height is correct; giving an estimate for his destination, and, if he wants it, requesting weather information.

But the FAA has an excellent service called Enroute Flight Advisory Service (EFAS). It is for weather only! One calls via radio; at the moment the frequency is 122.0 MHz, but more frequencies probably will be added because it's popular. It is called up using the name of the station first and then Flight Watch—so it might be: "Oakland Flight Watch this is Cessna 1737." The stations are listed in the Airman's Information Manual (AIM), Jeppesen, and elsewhere, but if you cannot find the list just call blind, "Flight Watch," and someone will answer.

This is the place to go to ask questions about weather. They are up-to-date and that's the only thing they work with— weather! Don't try filing flight plans and all that on EFAS; go to FSS. We can feel relaxed using it without that feeling that other traffic is being held up as we ask for weather.

A lot can be gained by just listening to the frequency as others ask for weather or give it. Often just listening will get us what we want without ever making a call.

A most important part of EFAS is what pilots tell them! They want to know what the man in the air is experiencing. It's useful to other pilots and meteorologists for weather and forecast updates. So EFAS—Flight Watch—is a three-way street: Listen, Ask, and Tell.

If a VFR pilot does all these things as he flies VFR, he will develop a facility that will be useful in instrument flying. Practicing and learning VFR will make passing the instrument test easier and, most important, will mean that when he starts to use his newly acquired instrument ticket he will be familiar with things and able to get more use, sooner and more safely, from his rating. There is a lot of value, present and future, in using many IFR procedures when flying VFR.

WITHOUT RADIO

The pilot without radio cannot enjoy the luxury of obtaining weather information in flight, so his chore is to get a good weather briefing before takeoff and implant a good picture of the situation in his mind. Then as he flies he watches clouds, wind, and visibility, being constantly wary and alert to action possibly differing from the forecast, ready to realize that something has started to happen that wasn't supposed to.

He will not have the luxury of always knowing what's going

on, but he may have more fun because the game is tougher to play and so the prize is more rewarding. He may, too, after he gets radio and an instrument rating, be a better weather pilot.

So now on to the pilot with an instrument rating—a piece of paper that doesn't necessarily mean he's a weather pilot because, at first, it's only a learner's permit.

About Keeping Proficient Flying Instruments

We may have an instrument rating, but unless we stay proficient the rating isn't really any good. How do we keep sharp?

Airlines, of course, check their pilots twice a year. The checks serve two purposes. One is to see if the pilot is flying instruments as well as he should. The other is to let him do things he never gets a chance to do in routine flying.

If an individual wants this kind of checking, there are places he can go and pay for it. Both check-flights and refresher courses are an excellent idea.

But aside from going to professionals there are ways we can stay at top level and test our own abilities.

PRACTICE

First of all, is the way we fly every day. What we should do is fly precisely and practice in good weather or bad.

This means holding altitudes and headings as closely as possible, flying precise airspeeds in climb and descent, and making corrections smoothly and exactly. If we do these things, they become a habit.

An important part of this is to use the ILS every time there's one on the runway we're landing on, good weather or bad. The more practice the better.

It's also interesting to shoot ILSs on a VFR basis occasionally, to see where the ILS takes you. One learns about bends in both localizer and glide slope and gets a chance to see what various altitudes and terrain clearances look like along the ILS. He can learn how far "off course" it's possible to be and still make a turn to get on the runway or, even more important, how close one must be to the localizer at minimum altitude in order to have enough room to turn and get on the runway.

Different ILSs act in different ways. They have their own characteristic bends. Also, glide slopes are not good all the way to the ground. Their signals become useless at some altitude near 100 feet. I've found that this altitude varies with individual glide slopes; some are good down to 80 feet, while others "come apart" at 150 feet.

By practicing approaches we can see where the glide slopes at runways we frequently use lose their accuracy. All this can be handy knowledge on some stormy night when we really have to shoot a tough one. On the routes I fly, I try to know each ILS at each airport and runway.

Proficiency is a matter of practice, and that is a matter of doing. Almost the only pilots I've ever seen who have trouble on instrument checks are the ones who get in the habit of doing only the necessary flying and no more. If it's VFR, they just go in and land. They don't make use of flying's opportunities. I've seen fellows have trouble and then when awakened by the jolt of a down check get to work and come back to top level proficiency. Generally, one jolt lasts a lifetime. It not only raises the ugly idea of being without work, but it's a real injury to one's pride.

SELF-CHECKING

We are also interested in how good we are, what sort of situation we can handle. Well, in the absence of a check pilot, let's be our own check pilot.

The first thing we need is someone to ride along who will look out for traffic and who can recover if the airplane gets out of control. Two pilot friends can ride for each other and learn a lot in the process.

Once in the air and at a good altitude, 4000 or above, put up the hood and cover the Horizon and Directional Gyro. Then do these exercises on the primary flight group: needle-ball, airspeed, and vertical speed.

The exercises should be done without any appreciable altitude loss, overshooting or undershooting altitudes when descent or climb is called for, or great airspeed fluctuations. Any indication that you don't have the airplane under control, that you are unable to keep it from stalling or exceeding its high speed limits, means instruction and practice are needed before you do any instrument flying.

So here are the test maneuvers, under the hood, primary flight group:

1. Straight and level for five minutes.
2. Climb at 300 feet per minute—airspeed 50 percent above stall. Climb like this for 1000 feet. Level off at an exact altitude. Then descend at 500 feet per minute, same airspeed, for 1000 feet and level off where you started the original climb.
3. Do a straight-ahead stall and recovery. A real stall! Not an approach to stall, but one that makes it shudder and shake and want to duck a wing.

4. One needle-width turn to right for 360 degrees. Maintain altitude within 50 feet. Roll into one needle-width turn to left for 360 degrees. Hold altitude.

5. Stall from 30-degree banked turn.

6. Establish three needle-width turn to left for 360 degrees. Keep turning and start descent allowing airspeed to build up to 30 percent above cruise IAS. (Don't exceed any limits.) Descend 1000 feet. Stop descent, level off, and come out of turn.

7. Following above immediately roll into three needle-width turn to right. Do one 360-degree turn and then climb in the turn at 70 percent above stall IAS for 1000 feet. Level off and stop turn.

If you can do all these things neatly and with precision and be master of the situation all the time, you are a pretty good instrument pilot.

WITH FULL INSTRUMENTS

Now uncover the DG and Horizon and fly with full panel and do the following exercises. They are to show the precision demanded when flying in a traffic area and executing an approach and pullup. They should be done with great exactness: airspeeds within a few knots, altitudes exact and not varying over 50 feet, headings hit on roll-out within three degrees.

1. Straight and level. Hold exact heading and IAS.

2. Reduce speed to 50 percent above stall. No altitude gain or loss.

3. Reduce speed to 40 percent above stall. Turn 45 degrees to right. No altitude change and hit new heading within 3 degrees on roll-out.

4. Maintain above speed and descend at 300 feet per minute. Turn left 170 degrees. Lower gear.
5. Level off, lower flaps to landing configuration, reduce to approach speed. Hold this for two minutes.
6. In above configuration descend at 500 feet per minute to a preselected altitude 1000 feet below.
7. At that altitude do a pull out. (Tune omni to new station and set in a different radial.) Turn left 90 degrees as you do. Establish climb speed and clean up gear and flaps. Climb 1000 feet and level off.

These exercise maneuvers are to check your sharpness. They aren't an instrument pilot's final exam. They are simply something to go back to now and then for an appraisal. The degree of precision obtained says how good you are.

The maneuvers on the primary group are really designed to make certain you can keep out of, or recover from, a spiral condition with increasing airspeed and recover from a stall.

One can vary these maneuvers and design more for himself. The basic thing in the full-panel maneuvers is to try and load oneself up with all the things that might be required and see if they can be done with the precision demanded by the ATC system and, of course, safety. This means one needs an ability to change speeds while climbing or descending, turn to headings, lower flaps and gear, adjust rates of climb or descent, and hit altitudes and headings exactly. There's also a requirement to remember these altitudes and headings and keep them after leveling off.

One could do all these things and toss in an extra job of consulting charts and worrying about carburetor heat as well as talking on the radio. It can be quite a load, and we'd better face the fact and be prepared to handle it.

We should check ourselves periodically—every few months. An honest self-appraisal is often more severe than a check pilot's. This isn't the place to kid ourselves; if we're not up to snuff, we should practice more or go back for help from a professional instructor, or both.

12 Thunderstorms and Flying Them

A THUNDERSTORM is a thunderstorm, and while they may come from different beginnings they all turn into the same thing. Once a pilot is in one he wants to know how to fly through it and come out safely. Even more important, however, is keeping out of it in the first place; and no matter how old or bold a pilot may be, his primary thoughts about thunderstorms are concerned with staying out of them.

To help stay out, or fly through if we must, we should know what a thunderstorm really is and the different backgrounds they come from.

WHAT ARE THEY?

Simply a cumulonimbus, CB, or thunderstorm, T, whichever one calls it: a concentrated mass of very unstable air in violent motion up, down, and sideways. It has strong gusty winds that generally extend to the ground and make landing or takeoff a wild experience. It has thick cloud, heavy rain, and sometimes hail. Electrical discharges occur frequently. Some Ts have tornadoes associated with them, but most don't. Tornadoes, however, never occur without thunderstorms. Tornadoes are associated with thunderstorms, but they do not always come out of the big, bunched-up cumulonimbus clouds with the highest tops. They often come from much smaller clouds that hang back in a line from the main storm center. So it's wise not to duck under or fly through clouds close to a thunderstorm just because they look smaller and have lower tops.

To make a thunderstorm we need conditionally unstable air. What's that and who cares? Well, it's air that is stable as long as it doesn't condense . . . that's the condition. When it does condense, the release of heat in the condensation process makes the air warmer and it wants to go upward. Then more moisture condenses, more heat is released, and it goes up more . . . it's almost like perpetual motion and the energy released is tremendous; a nuclear bomb is a weakling by comparison.

The normal places we go for weather information cannot tell us if an air mass is conditionally unstable or not. That is learned from upper air soundings taken in many places over the country and analyzed by a computer at the National Meteorological Center (NMC), Suitland, Maryland. But if a weather forecast calls for thunderstorms, it means the air is conditionally unstable or will be, and that's all we need.

WHAT IS TOUGH ABOUT A THUNDERSTORM?

What really bothers us most in a thunderstorm is the turbulence. Lightning discharges are a minor danger and few if any airplanes have been knocked down by lightning. In recent years a question has come up of flame going up the fuel vent and then igniting a near-empty tank loaded with explosive vapors, and there is strong evidence that this might have happened. Redesign of the vent system, plus a method of discharging inert gases into the offending vent system when flame begins to move, shows promise of correcting this situation. But as for lightning hitting a plane and knocking it down, the evidence says it's doubtful. I spent a number of years looking for and flying through thunderstorms in research work and heard and saw a lot of lightning. I experienced numerous electrical discharges where the aircraft was one end of the lightning cycle ... four in one day ... and the worst damage was holes about the size of a half dollar burnt on the trailing edge of control surfaces. Bigger holes

Damage to a Boeing 707 from electrical discharge, often called a lightning strike. The point of discharge is small and near where fuselage and radome meet. Large, torn pieces were probably ripped out by airstream. None of this affected the airplane's flying ability.

have occurred, but they generally aren't caused by the lightning. When I had a discharge in a Boeing 707 out of London, England, a three-foot-long piece of the black radar nose was torn out. The lightning actually made a much smaller hole, but the high-speed air from our forward motion got inside the nose and then tore out the bigger piece. There wasn't any special problem, but I dumped fuel and returned to have it fixed before crossing the Atlantic.

TORNADOES

Tornadoes, of course, would be catastrophic if an airplane flew into one. The winds are unbelievably violent. But tornadoes can generally be seen if one is outside cloud, flying VFR. The big hazard is sneaking under thunderstorms at night or down low in and out of cloud. There is always the chance that while you are sneaking under a black cloud a spout may form from it.

I was lucky one night, in a DC3, flying from Chicago to Indianapolis while trying to squirm my way through a line of thunderstorms. I was flying low, as we often did in pre-radar days, trying to stay "contact." Ahead I saw a strong glow which turned out to be a large field on fire, probably started by lightning. But the shock was to see this fire lighting the sky enough to show a tornado with its scary funnel right on my course! I went around it, somewhat shaken, but what if that handy fire hadn't lit up the sky! It's always useful to have a little luck around.

It does seem possible that an altitude above the base of the thunderstorms, perhaps 8000 feet or more, may be above the tornado. It will be a ride to remember, but the violently

What hail can do to an airplane. But it flew! (NOAA PHOTO)

destructive part of the tornado may be missed. But best, by far, is to avoid the entire area.

A friend of mine passed over the Pittsburgh, Pennsylvania, radio-range station in a DC3 many years ago at 8000 feet while a tornado was going by on the ground at almost exactly the same place and time. He didn't have a bad ride and flew on to Washington, D.C. There have been numerous cases of this sort, so there's more than just theoretical evidence that the violence of a tornado is in the low levels only. Part of the reason for my friend's good ride was that he probably missed flying through a cell. It is evident that tornadoes go up into the cells, so if your radar leads you around the cells, you'll probably miss the tornadoes, too. These aren't solid facts, however, and some tornadoes come out of lower clouds behind thunderstorms.

One criterion, developed by Dan Sowa of Northwest Orient Airlines, is that if the reported radar top of the thunderstorm pushes up above the tropopause by 10,000 feet, the storm has tornado potential. One can get thunderstorm and tropopause tops from reports via NWS, FSS, and Flight Watch.

HAIL

Hail is always a problem, but compared to the numbers of airplanes flying in and around thunderstorms, not many are hit by hail and there isn't any record of an aircraft actually being knocked down by it—although a few have been pretty well beaten up. There is evidence that big hail can damage a jet engine so it loses its effective power. Apparently one catastrophic forced landing resulted from it.

Most hail comes out of the big overhang cloud that is downwind of a thunderstorm; if it's visible, this is the place to stay away from, even if it looks clear and pleasant there. You might get right under it when the hail dumps out. Thunderstorms demand a wide, respectful berth!

Heavy rain can cause engine problems in the form of carburetor icing and all precautions must be taken. Sometimes the rain seems so heavy that the engine should simply drown out, but evidence does not suggest that there is any danger of this happening.

THE BAD PART

The big problem, as we said, is turbulence sufficiently severe to make keeping control of the airplane difficult. The turbulence may be severe enough to tear an airplane apart, but it is most likely that a structural failure will follow loss of control and subsequent high airspeeds that may reduce the airplane's structural integrity.

Airplanes are pretty tough and, if flown properly, will probably traverse a thunderstorm safely. This isn't meant, in any way, to suggest that pilots fly through thunderstorms on purpose . . . it is and always will be best to stay out of them. There is some evidence that if certain severe thunderstorms are encountered at the point of their most potent growth, it may not be possible to maintain control or keep the airplane in one piece.

Landing with thunderstorms close to the airport is a hazardous operation. Most landing shear accidents have occurred with thunderstorms close, and some not so close, to the airport. A thunderstorm may have a gust of air well ahead of it—called a gust front—that will make a large, dramatic wind shift. If there is a potent thunderstorm, squall line or front within 15 miles—20 is better—of the runway, extreme caution must be used and all the surface wind information possible obtained. Best by far is to avoid landing until the condition has moved off.

THEIR LIFE CYCLE

Thunderstorms, like everything else, are born, live, and die. They do it faster than one might think, but as one dies another forms.

You do not see all this when approaching an area of thunderstorms. Large clouds pile up in great masses, and unless there's just one lone thunderhead of an air-mass or orographic type, there will be high, intermediate, and low clouds as well as the cumulus.

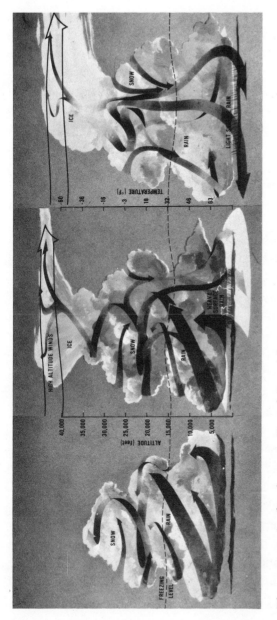

Three stages of a thunderstorm. From left to right:

1. *The early or building stage. Lots of updrafts and rough.*
2. *The mature stage. Heavy rain, possibly hail; up- and downdrafts; chopped-up, rough air; and strong surface gusts ahead of the storm.*
3. *The drying stage. Mostly downdrafts. Rain diminishing and not very rough. But it still looks that way.* (NOAA PHOTO)

All these clouds aren't wild and rough; most of them are not and contain only light to moderate turbulence. But buried in those clouds are thunderstorm cells, anywhere from a few thousand feet to six miles or so across. In these cells things are rough and wild. They are the heart of a thunderstorm.

The storm may have started as a snappy, jolty little cumulus that formed and started to grow. It continued to grow and finally became so high that it passed the freezing level; there it stopped being a simple cumulus and became a thunderstorm.

This is the cumulus stage, and during it the cloud is almost all updrafts, some going up 3000 feet a minute; there isn't any anvil or rain.

When I was doing weather research, we measured electrical fields and flew a lot of CU and thunderstorms doing it. One early afternoon, over the Pocono Mountains of Pennsylvania, I flew into a large growing CU that was going through about 14,000 feet, but didn't show any signs of being a thunderstorm. It turned out to be one of the roughest rides we ever experienced, however, and the electrical field strengths were tremendous. Which says it doesn't have to be a full-blown thunderstorm, cracking lightning, to be tough.

The way to study a growing CU is by careful scrutiny of the very top part of the cloud, the shreds and pieces. If one can see a spilling, growing action, and if it's "busy," there's lots of action inside and it's growing. Any indication of a small, flat, thin cloud around the edges, near the top, is strong indication that this CU is going to become a thunderstorm.

The mature stage follows the cumulus. It begins when rain commences. The cloud develops downdrafts as well as updrafts along with heavy rain. It's still very rough because of the added conflict of air moving up and down.

After the big anvil has formed off the top, the thunderstorm enters the dying stage, the wild currents decrease, and it calms down, although it still looks pretty bad.

That's the way thunderstorms come and go, but from a cockpit we want to know more practical things: where are the tops, the bases, how far through, what's in there, and very important, what kicked them off in the first place, because if we know that, we have a start in knowing how to combat the particular storm we're faced with.

There are three basic ways thunderstorms are created:

1. By heating.
2. By a front.
3. By orographic influence such as air moving up a mountainside or sloping terrain.

All these, or any two, can unite and produce thunderstorms faster and more violent than usual.

When a thunderstorm occurs because of air-mass moisture and instability, the air mass will be heated faster and rise to its condensation level quicker on the sunny side of a hill, where glider pilots would look for thermals; a flow of wind up the mountain will also add impetus to the process.

A front shoving air aloft on its own will do it faster and with more umpf if that front is climbing up sloping terrain.

So hills and mountains make thunderstorms tougher, as they do other weather, and the rule of extra care in mountains applies again.

Thunderstorms are graded on reports with a system of numbers from 1 to 6, at present. Number 1 is light, and 6 severe. They are also rated light to severe on weather reports by − or + like other weather.

This rating system can be misleading in that pilots sometimes think an area of class 2, for example, isn't too bad for flight. But this thinking can squeeze a person into trouble because thunderstorm intensities can change unpredictably and what may be called a 1, or light, may well turn out to be severe, or a 6! As alarming as it may sound, we'll do better remembering that all thunderstorms are bad, some are just worse than others!

A CLUE

When the weather services report actual thunderstorms, they give the direction from which the storm is moving and its velocity. This velocity should not be confused with the general speed of the weather system, such as a cold front. The speed is strictly of the individual storm. There is definite correlation between this rate and the storm's severity. If it's over 20 kts the storm will be strong, and if over 30 kts it will be severe. In one of our worst airline thunderstorm accidents the cells were traveling more than 60 kts.

If a rain shower—not a thunderstorm—is moving over 50 kts, it will have strong turbulence and deserve caution.

THE DIFFERENT KINDS

Now what do the different thunderstorms mean to us?

First, air-mass thunderstorms. They are generally scattered, and theoretically we see them and wander around getting where we are going without flying through any storm.

We see this in the Far West where thunderstorms occur in classic form. The visibility in the semiarid areas is excellent, and if anything less than 50 miles is reported one looks at it with suspicion.

Here it is easy to waltz gracefully around the big cumulonimbus, watching its dark, foreboding rain and flicks of lightning from a safe and interesting position. Away from these regions of unlimited visibility, however, air-mass thunderstorms become more of a problem as we fly in Midwest and Eastern summer haze.

If we are working our way through air-mass thunderstorms in reduced visibility, the place to be is on top of the haze level where we can see the thunderheads. It's worth repeating that when doing this it is important to make certain that clouds do not creep in under us and that we keep track of the general weather situation.

Air-mass thunderstorms do occasionally line up in a sort of fake front if the condition favoring the thunderstorms is strong. This generally occurs in the late afternoon.

HOW HIGH

Here we should talk about thunderstorm tops. They are high—35,000 to 80,000 feet, depending on the part of the world. Except in rare situations, any CB worthy of the name doesn't stop at 25,000 feet. Almost any fully developed thunderstorm will keep right on climbing until it pokes up into the stratosphere.

The tops are generally lower in northern latitudes, and so is the stratosphere. I've flown over the top of thunderstorms at 33,000 feet above the North Atlantic, but in southern Italy I've flown next to thunderstorms at 35,000 feet that looked as far above me as I was above the ground.

Thunderstorms' tops don't extend above the stratosphere's bottom because the stratosphere is an inversion: the air is warmer than the rising storm air, and so it's stable and shuts

This picture of an isolated thunderstorm of the type found in the western United States was taken from 32,000 feet. The top is still a long way up. An end run around the left looks good, but not too close under that overhang. Generally a run around the upwind side is better, but watch for the overhang from another cell farther upwind.

off a thunderstorm even though its momentum may drive it up into the stratosphere a thousand feet or so.

In the USA the tops are rarely below 35,000 feet. It's obvious that small aircraft don't try to top thunderstorms and big airplanes don't too often either. There are problems for both kinds.

If a person does try to top a thunderstorm, or a line of

them, he should be prepared for turbulence even in the clear; rough air often extends above the storm's top and if an airplane is staggering along at a high angle of attack, struggling to get on top or stay there, it may be so close to stall that a bump will cause a stall; the aircraft may then wind up out of control or in a recovery dive inside the storm—a wild combination. Stall recovery at altitude isn't just a simple dip of the nose. It takes a big altitude loss to get back flying.

It's impossible to overstress the danger of trying to top storms, and even going over them with apparently good speed and control is a more risky flying condition than one may realize.

WATCH THE CLOUD LAYERS

It's not difficult to get in a bad situation trying to stay on top of lower clouds around storms. Let us say we are on top of the haze level. We see numerous cumulus (CU) poking up through; they look like bunchy, thick cauliflowers. They are easy to wander around, but finally they seem to be connected by lower clouds and the top of these lower clouds approaches our cruising level. This can happen, depending on the stage of development, anywhere from 8000 feet to more than 30,000.

Almost without realizing it we've been climbing to stay on top and our airplane is grunting to do the job. We are being suckered in, as the blunt expression goes.

Big cumulus tower to either side of our course. Ahead is a lower spot with bunched up CU. We head for the lower place and hope we'll get over and beyond before it builds up to our level.

THEY GROW FAST

Something we should realize, however, is that those CU can build very fast and turn into thunderstorms. I watched one build in Texas. At 11:30 it was a pleasant looking CU, at 11:45 it was a big, solid, bunched-up-looking CU, and at 12:00 it towered to tremendous heights, its bottom was black, rain poured down, and lightning crashed to the ground!

When one stops to consider that thunderstorms in the CU stage can have updrafts of as much as 8000 feet per minute, with 2000 to 3000 feet per minute fairly routine, it's obvious that a growing thunderstorm can outclimb an airplane of average performance.

So as we sneak through that "low" spot we should realize there's an excellent chance the "low" spot will come up to us, around us, and envelop us. One cannot try to top growing CU unless willing, able, and prepared to fly through a thunderstorm, because that's what it may finally come to!

A pilot able to fly only VFR simply shouldn't get in this position. While he may be on top of the haze level to spot CU buildups easier and fly around them, he just hasn't any business flying on top of bunched-together masses of CU even if the tops look low. When the CU surround him and the only path ahead is over some, or through low spots between them, then his only option is to turn around and/or land.

The instrument pilot who feels he can flirt with growing tops, trying to slide over the "saddles" between them, had best review his capabilities first and remember that no matter how experienced he is, a stalled airplane in a thunderstorm is in a desperate situation, and the cleaner the airplane, the more desperate.

Cumulus building. The low spot between will build, too, creating a situation in which one will have to climb higher and higher to stay on top. Thunderstorms will probably develop. For VFR pilots it's a situation that requires working calmly on getting down and out of it all. For IFR it may mean an instrument clearance or climbing on top if the airplane can get high enough—15,000 feet plus—and then wiggling between thunderstorms.

As we said, if this kind of flying is going to be done, then one must be willing and able to fly in a thunderstorm.

What's willing and able? First, one must be a competent instrument pilot who can fly the airplane well in heavy turbulence. Then, one must have instruments enough, good engine-heat capability, and radar.

WHAT'S INSIDE ALL THOSE CLOUDS

Let's go back to our line of towering CU with the lower spots between. As time goes on the lower spots will be higher and finally a solid line of massive CUs and CBs cuts across our course; they may be lined-up air-mass thunderstorms in the afternoon, or a genuine front. Either way it looks impressive and has the appearance of one big line of thunderstorms.

If we could really see inside that line we'd find a row of heavy clouds with numerous thunderstorm cells inside. If we barged through and missed the cells we'd get light-to-moderate turbulence and light-to-moderate rain and, on a chance basis, hail of any size.

If we barged through and encountered a cell we'd have a wild ride, and if it's a cell just growing, young, and vigorous, we'd have a terrible ride.

This has often been demonstrated by two airplanes going through a husky cold front on the same airway, same altitude, and perhaps only a half hour apart. One pilot says he had a horrible ride, terrified all the way. The second pilot looks a little questioning and may think the first pilot an alarmist because he himself had a pretty decent ride. The difference in rides wasn't due to a difference in fortitude between the pilots; it was simply that one missed the cells and the other didn't.

RADAR

This brings us to radar. With it we have an excellent way of seeing the cells and avoiding them, although it takes learning, experience, and radar equipment in good condition. But what we want to say now is that we should never be on instruments when thunderstorms are nearby unless we have radar and know how to use it.

Radar is a tremendous aid in thunderstorm avoidance. What it does is let you see cells on a scope that you cannot see by eye because of other clouds and darkness. But note, carefully, that radar doesn't help you fly through thunderstorms. With or without radar, once in a thunderstorm it's a wild, undesirable ride.

Look at the thunderstorms by radar as far in advance as possible and then attempt to fly around the entire area, or the biggest area of cell congestion, rather than getting in close and using radar to squirm between cells. A good look and planning in advance can mean missing the entire mess.

There are more advantages to reading the picture in advance. For one, it allows time to tell ATC what kind of headings are desired to circumnavigate the area well in advance of arriving there. ATC will be in a much better position to cooperate and give the needed course changes.

Another point is that once in close to the cells, all will not be easy. The radar does not tell you what's behind one group of cells, and so we get surprises as we round a cell and see a new line ahead of us. Cells tend to generate close to old cells, so being in an area of cells, and close to them, means that more will develop and getting through it all without getting roughed up, even with radar, isn't going to be easy.

Being in close to the cells doesn't give much room to avoid them by the respectable distances that good procedures dictate—we'll talk about that in a moment. But all in all, the best way to use radar is to duck the entire area.

Remembering, then, that radar is a device to help you miss

cells, it's apparent that when radar fails to show storms, it's a useless device.

Radar is fallible; there are times when it doesn't do the job and we'll talk about them here.

Before we do, however, let's make a point clear: If you are going to enter a thunderstorm area depending on radar to lead you through without getting in a cell, you had better be able and willing to fly through a cell if you have to.

Radar will take you places you'd never go without it. Suppose you get into this forbidding place and the radar fails; now what? You are either lucky and fly through without running into a cell, or you bang into one. If you hit one, let's hope you and your airplane can handle it!

Radar's important feature is that it reflects from rain; but at high altitudes the cloud doesn't have rain, but snow and ice crystals instead. The radar beam doesn't reflect as well from this frozen stuff, and the cell becomes difficult to see on the scope.

The way to try to overcome this is to point the antenna down toward the rain area for a picture. The trouble is that the radar then bounces off the ground too and the picture becomes confused, and individual storms difficult to see. It takes a lot of experience to read a scope under these conditions, and even with experience it doesn't always work.

Radar maintenance problems are not always just a simple matter of whether it works or not. There are degrees of radar maintenance, and this makes thunderstorm reading difficult and less clear.

It is necessary to get all the information possible from the manufacturer and learn to recognize when radar equipment is not putting out its best. There are various ways of doing this; one is to note how much "gain" the tube requires for a picture. If the knob has to be twisted farther toward high "volume" than normal, things are getting weak and it's time to have the set checked.

Let's list radar's limitations:

1. Total failure.
2. Partial failure.
3. Set deterioration.
4. Poor reflection from frozen particles.
5. Pilot ability to tune and read.
6. Inability of set to show any but rain areas.
7. Difficulty of reading in mountainous terrain.

With these listed we can further list two rules:

1. Even with radar, don't go into cloud areas containing embedded thunderstorms if you can avoid them.
2. If in cloud or darkness and avoiding thunderstorms with radar, give them as wide a berth as possible. Don't just skim around the edge of cells.

There are a lot of techniques in using radar. It is best to learn them at school. There are courses and excellent movies available to help gain this knowledge, and anyone who has spent the large sum necessary to acquire radar had best invest a little more for schooling.

One important point is the general guide most airlines use in determining how to miss cells. It's based on temperature and altitude. They say to miss cells by the following amounts:

When the temperature is above freezing—five miles.
Temperature below freezing—ten miles.
At altitudes above 25,000 feet—twenty miles.

These are minimums, and it's better to give the cells a wider berth if possible; 20 miles at all times is best.

The blow-off clouds on the downwind side of a thunderstorm is the place to stay away from; that's where the hail is, and often clear-air turbulence as well. In a confused mass of cells and, of course, at night, we cannot see this area. The blow-off drifts a long way downstream, so a rule is: Avoid a cell, on the downwind side, by one nautical mile for each one kt of wind at the flight level. That may be a lot of miles, but it's worth them all.

Best is to go around on the upwind side and avoid the blow-off area. But one should not just skim by the cell on the upwind side, because there is wind shear in the area caused by a speeding up of the wind close to the cell—and that's turbulence. So on the upwind side, miss the cell by at least 10 miles even if flying in the clear.

MORE ABOUT AIR-MASS THUNDERSTORMS

We were talking about air-mass thunderstorms; it's important to realize that these come in several different kinds.

The simplest is the convective kind. They are an eventual development from a thermal source, a hot spot on the earth. Obviously, when the sun goes down the heating does too, and the shower dissipates. If it doesn't then the storm isn't just a convective storm; something else is creating it. The real thermal convective storm may take a while to die and so doesn't disappear as soon as the sun goes down; a big bunch of clouds hangs into the darkness with an occasional flick of lightning, but it's on its way out and one can see and feel that it isn't holding or increasing its strength.

Thermal convective storms are easy to run around when they are scattered or isolated. Sometimes, however, they aren't so scattered. Whether they are scattered or broken depends on how much moisture is in the air mass and how unstable it is. This is pretty difficult for a pilot to judge, except that forecasts give an idea simply by the number of storms they call for. Our old friend the temperature/dewpoint spread is another indicator; if the dewpoint is high, the air is moist and more likely to produce storms. You can "feel" this too, and since weather appraisal is inexact to begin with, one shouldn't laugh off old sayings like those about rheumatism that hurts when its going to rain. On a hot, muggy summer day any weather-conscious person can sniff the likelihood of storms.

Man is still basically an animal, and despite years of sophisticated city living and separation from the primitive state he has a feel for weather; he feels it in his bones, as the saying goes. Naturally we don't use instinct for our basic weather judgment—science is the important thing—but that old instinct is worth using, particularly when it raises suspicions that all is not as good as a forecast may say it will be.

A CLOUD BASE HINT

Incidentally, temperature/dewpoint can help in finding the base of cumulus clouds, or how high one will have to climb to get up out of the lower-level convection. Take the temperature/dewpoint difference (in Fahrenheit) and multiply it by two. The answer is the altitude in hundreds of feet. For example: temperature 80, dewpoint 60. The difference is 20. This times 2 is 40. The base of the cumulus will be about 4000 feet.

OTHER AIR-MASS THUNDERSTORMS

To sum up, air-mass thunderstorms are generally isolated and occur in the afternoon when heating has had a chance to work on them. They can also occur in overrunning air, not necessarily from a warm front, but from slopes acting like a warm front. This will carry thunderstorms into the night, but probably their base will be high.

Air-mass thunderstorms can bunch up and look like a front, but they can generally be flown around. In the parts of the world where visibility is good, such as in the far western USA, this is easy. In the more moist sections, where haze reduces visibility, it's more difficult; a pilot is best on top of the haze level where he can see the CU buildups. It's important, of course, not to get trapped on top when doing this.

It is also important not to try to top thunderstorms, unless with "super" equipment such as powerful jets with 50,000-foot-plus capabilities.

It's dangerous to sneak around underneath high overhanging anvil clouds because that's where hail falls from. It falls more outside the storm than inside it.

It is dangerous to sneak under thunderstorms where visibility is reduced, and there's a chance of flying abruptly into clouds, rain, and turbulence. Mountains may be buried in the haze too, and TV towers. Also, if one runs into turbulence down very low he's in a more hazardous situation than he would be up a bit higher where he has room to wrestle the rough air.

While trying to get past a line of thunderstorms, down low, we sometimes see the clear-cut black edge of the storm's base and beyond, happily, the pleasant sight of unlimited visibility and clear sky. We become eager to pass this line, and go lower to stay "contact" and fly under. The nose is pushed down and the airspeed goes up, so that we're apt to start under the storm's edge going fast. But under this edge will be some really rough air with hard, sharp-edged gusts, and high speed will be dangerous. So patience is needed to keep slowed to turbulence-penetration speed as we sneak under—and we should realize that there's tough flying yet to be done before our peaceful sky is won.

In early airline days a DC3 was flown under such an area in a flat but high-speed dive by an overeager pilot. The jolts resulted in injury to ten passengers and some wrinkles in the airplane!

FRONTAL THUNDERSTORMS

We visualize frontal thunderstorms as part of a dark, vicious cold front moving across the countryside with high winds, pouring rain, and much lightning, and that's a pretty accurate picture. But there are warm-front thunderstorms too and they are different.

What's the difference? Well, it's mostly in the height of the storm's base. A cold front's storm base will be close to the ground, with no room to fly under without getting the full force of the storm. But a warm front's storms, except where the front meets the ground, are generally aloft. How far aloft will depend on the frontal slope. I've seen thunderstorm bases as high as 16,000 feet, which is unusual. But I've seen lots at 4000 and 6000 feet. Between Chicago and Kansas City, for instance, I've often observed thunderstorms reported over most of the route. But by examining actual weather reports closely one could see that the bases were high, which indicated a warm-front condition. A flight at 2000 feet put the airplane in smooth air although there were clouds, stratus from falling rain, and lots of rain and frightening lightning all the way.

THE SURFACE WIND TELLS

How do we learn it's that kind of condition? First, as always, is a look at the big picture. Where are the fronts and what kind are they? Next, a careful study is made of the surface-weather reports with particular attention to surface wind. If the storm bases are high, the surface winds will not be strong and gusty at stations reporting thunderstorms. The wind will pretty much stay normal for the general circulation. This is the best clue that the storm is a high-base one.

A warm front will also have storms over a large area with extensive cloud cover. A cold front, in contrast, will show up as a line of thunderstorms and can be seen on surface weather reports by the strong ground winds and a dramatic wind shift as it passes a station.

A warm front is of special interest to the turboprop and light pressurized twin, because their normal altitudes may put them near the warm front surface aloft where the thunderstorms are wildest. The pure jet will get on top of all this, or up where it's easy to dodge around, and the little guy will be lower and probably under the thunderstorm bases.

HOW TO TELL A FRONT'S TOUGHNESS

The intensity of a cold front can be learned by studying the sharpness of the wind angle ahead of and behind the front. If the wind in advance of the front, for example, is from about 210 degrees and the winds behind the front are from 340 degrees, we can bet the front will be a wild one. Also, the larger a low's warm sector, the more violent a front. The bigger the temperature difference between the warm and cold side of the front, the tougher it will be.

It is also important to check the speed of a cold front by looking back over the sequences to see how fast it has been moving. If it speeds up you can bet it's getting stronger.

And, of course, there is our old friend orographic influence; if the front is moving up sloping terrain, that too will make it worse.

PREFRONTAL SQUALL LINES

Cold fronts have a phenomenon associated with them that complicates the picture: the prefrontal line squall. This is a line of thunderstorms that breaks out 100 to 300 miles ahead of a cold front. They occur mostly in the afternoon, when heating adds to the zip of instability, along with the lifting effect of convergence in advance of the front. All this is difficult for the day-to-day pilot to analyze; he must simply be suspicious, always, of a prefrontal line squall when a cold front is on the move.

This squall line ahead of the front is where the toughest thunderstorms are found. It is where most tornadoes occur. It is where most hail is found. It is a good thing to avoid!

As a prefrontal line squall develops and intensifies, the cold front behind it tends to diminish and sometimes can be difficult to find on the weather sequences. As evening approaches, the prefrontal squall line will diminish and then the real cold front will make itself known again. What this means is that if we negotiate a prefrontal squall line we may think we have finished with its violence. It may turn out, however, that 300 miles along, as night approaches and we fly toward the dark, our cold front has perked up and we find ourselves facing another line of thunderstorms.

A good rule of thumb is that prefrontal line squalls are

generally not active within 150 miles of a low center, nor more than 500 miles out from the center. They show up in an area 50 to 300 miles ahead of a front and roughly parallel to it. Sometimes the breaks in a prefrontal squall line are larger than in a solid cold-front line, but this is a flimsy thing to depend on.

SOME RULES

The same rules apply to cold fronts and prefrontal line squalls:

1. Flight on instruments should not be attempted unless equipped with radar in good working order, and with a knowledge of how to interpret what the radar shows.
2. Low flight underneath should not be attempted because of extreme turbulence and the possibility of heavy hail.
3. The best procedure is to fly well around the line of storms or land and wait for it to pass. Landing and waiting isn't too great an inconvenience. First of all, these sharp lines of weather pass fairly quickly. Secondly, in a year of flying, most pilots don't encounter such conditions more than a few times. Of the total year's time flown, that rare couple of hours sitting relaxed on the ground is well worth it.

IF WE FLY THROUGH

Now, of course, comes the time when, for some hard to imagine reason, we barge ahead determined to go through that cold front. Where should we take it on?

Down low isn't good, as we've said before. Another place

to avoid is the area where temperatures are 20 to 35 degrees F. Research has shown that this is the area where most lightning discharges occur. It's also an area where carburetor ice will be at its maximum.

A logical place to go through is just above the haze level, at the inversion. Here there is a minimum effect from heating, and that may be a place where the violence is a little less—and we emphasize *little*.

We can visualize flying toward a front. The sky is hazy and visibility poor in the hot mugginess. Cumulus clouds dot the sky. To avoid them, see better, and be in smooth air, we climb above the haze level. As we climb, the temperature decreases, but just as we come up out of the haze to where the visibility is unlimited, the temperature rises. That's the inversion.

Now we can see the CU sticking up through the haze. We are over Ohio flying at 10,000 feet. We should be listening to each weather broadcast on our route, noting which stations have southerly or northerly winds as an indication of the front's location. The wind velocity, the abruptness of the wind-direction shift, and the temperature difference across the front are checked as well as the presence of thunderstorms at various stations en route.

Flying further west we note that the tops are climbing. We are at 12,000 feet and ahead is a line of CU from one end of the horizon to the other. That's our front.

This line of CU reaches very high. At first we have hopes of topping it, but experience says we're kidding ourselves.

The line has higher places with anvil tops. These are the cells of the storms. With radar we can see them and begin to plan a route through. Without radar, we'll make an educated guess.

Close to the front there will be a towering cloud wall ex-

A great picture of a line of thunderstorms along a cold front. The "open" places between the storms are very high and will tend to close up as other cells develop. This isn't a front to try VFR. It's an instrument flying job, with radar equipment required. (NOAA PHOTO)

tending to neck-stretching heights. Below will be stratus and low, boiling roll clouds. At our level, just above the haze, we'll see some stratus and layers. The wall of the front is a solid mass, bulbous in places; its color is eerie, anything from misty white to black yellowish-green, depending on the sun's position in relation to the front.

We study the complex picture before us and try to decide how to take it on. We don't want to descend and try to get under that roll cloud, because the air is wildly rough right

there, probably the roughest part of the storm. It's tearing up in tremendous currents and it's chopped up and inconsistent. Close to the ground the wind over the surface adds to the thunderstorm's normal drafts. Progressing through this roll area we'd run into heavy rain and zero visibility. The rain might affect our airspeed and there are also localized pressure changes that will affect the altimeter's accuracy (no one can tell you how much for certain, but some knowledgeable researchers think it may be as much as 1000 feet!). So let's forget low down as a place to go through.

We look up, way up. We've probably studied this area for a long time as we approached the front. What we are looking for is the lowest point in the top of the long line of clouds, a saddle between towering CBs. When we find it our first tendency is to climb right up there and slide over between the CBs. But as we've said, this is risky business, because just as we are approaching the saddle, the clouds will build up in front of us and suddenly there isn't any saddle, or it's much higher than it looks. The risk, to repeat, is that we'll find ourselves struggling high up at a mushing, high angle of attack ready for a stall. Unless we can zip over this saddle between CBs with good speed and control, we shouldn't fool with it.

AT NIGHT

Finding the right place at night is more difficult, but the flashes from lightning help. Approaching the storm area we study the cloud structure in the brief lighted-up periods as flashes occur. Often we can spot the individual storms by the concentrations of lightning. It is well to do this from as far away from the cloud mass as possible because close to the cloud the origin of the lightning becomes confused as it reflects and lights up all the clouds.

Lightning can be seen from a long way off at night, and storms that may seem on top of one can still be 50 or 100 miles away.

Close to the storm area, when looking out, care must be taken that a nearby lightning stroke doesn't temporarily blind one. When you're that close in, it's time to have the cockpit lights up bright and the decision made as to where to penetrate the mess.

WHERE TO BORE IN

We've decided that we cannot top the line anywhere. So we've picked a point under the lowest place in the top of the line confronting us. We're above the haze level and eye to eye with the solar plexus of the storm.

Before barging in we ought to study the color of the precipitation ahead. If it's white and fuzzy, with a gossamer look, it may be hail. If so, then we'd better wander up and down the line some more and find another spot that looks darker, like rain, and not whitish, like hail.

Let's restate, before we go in, that we don't recommend it.

We have our spot picked and are about to go in. Now the problem changes from weather judgment to flying technique.

HOW TO FLY IT

First we prepare the airplane. We pull down on the safety belt and put on shoulder harness if we have it—and we should. We make certain loose items are fastened down and passengers well strapped in.

Put on heat, for the pitot, carburetor, or jet inlets. Now establish the airplane's best rough-air penetration speed, and note what the power settings are for that speed and where the stabilizer-trim control is set. Note the airplane's heading.

But first let's talk a little about the penetration speed. Basically it's the slowest speed possible without danger of stalling. We want it slow so that the airplane will be in its best condition to take heavy gust loads. Too slow, however, has proved to be almost more dangerous than too fast. As we've said before, when airplanes come apart in severe turbulence the problem begins with loss of control, and an easy way to lose control is to stall, or to get a wing down and into a spiral dive.

Clean, swept-wing airplanes are worse in this respect than slower, straight-wing airplanes. If they lose control they gain speed quickly. A swept-wing airplane isn't as straightforward in its elevator-control system as a conventional airplane because, often, the amount of elevator control available is dependent on the stabilizer setting. On certain jets one runs out of elevator control unless the stabilizer setting is correct. In effect the stabilizer is the elevator and the elevator is like a trim tab. A more conventional airplane doesn't run out of control; it just gets harder to pull or push.

Sometimes the adjustable jet stabilizer cannot be moved at very high speeds, such as we might get in a dive, and then we're in a real mess. It may be necessary first to push the control in the opposite direction to unload the stabilizer-trim mechanism so one can trim, and then pull in the direction needed. It's a trick requiring thought and coolness when plunging toward the earth!

It's best not to fly too slowly with any airplane. Most aircraft manuals recommend the proper speed for turbulence and it will be found to be something like 60 percent above stall. A 100-mile-per-hour stalling airplane may have a rough airspeed of 160 IAS. Jets may have higher speeds, but again, it's best to consult the airplane manual.

The airspeed we've picked is important, but equally important is how we fly it. What we do not want to do is chase speed with the elevator control and the airspeed indicator. The key to rough-air flying is to maintain a level attitude and do it by flying the Artificial Horizon. Note where level position on the horizon bar is with the airplane stabilized. Then keep it there. We do this with gentle pressures and not big pushes and heaves. Airspeed will vary, often wildly, and if we attempt to chase it by pushing or pulling we'll soon get into some pretty wild attitudes. The trim and power settings should not be moved to any large degree.

The proper pitch attitude, power setting, and stabilizer position are important to know in advance so one can set these up quickly and not fight an out-of-trim airplane while trying to settle down to the proper airspeed and power setting.

These settings should be learned, as we've said, on a clear day by trying them for various altitudes and weight conditions. But the best insurance is to get down to turbulence-penetration speed, power setting, trim, and attitude before one enters a turbulent area. Of course, there are times when we hit turbulence unexpectedly; that's why the clear air experimentation is useful to give us guidelines to shoot for when things are suddenly wild!

We have to face the fact that it is difficult, often impossible, and generally undesirable to maintain a given altitude when flying through the business part of a thunderstorm, ATC or no. Of course we try to maintain the proper altitude, but only as long as it doesn't require appreciable power or attitude changes. Too diligent an attempt to maintain altitude will result in wild attitudes and/or power changes which lead down the path to loss of control.

ARE WE SCARED?

A thunderstorm line is a pretty awesome sight. It's scary and any honest pilot, no matter how experienced, will tell you so. This fear is a factor to consider.

We've mentioned that scared or not a weather-flying pilot has to control his emotions and keep his imagination subdued or used only to advantage. The pilot has to fly carefully and thoughtfully even though his knees may tremble.

The toughest place to keep this fear under control is a thunderstorm. It is dark, it is turbulent, rain comes down in a deluge, lightning flashes close, you can hear thunder sometimes, and occasionally ozone from a nearby lightning passage can be smelled. A lightning discharge may make a brilliant flash and loud bang as it goes off the airplane. An irritating hashy noise may be tearing at our eardrums from static in the radio. It's a hell of a place for a man to be! It takes a strong will to say, "I'll watch the horizon and keep it level. I'll hold that heading and it will all turn out okay." But that's it, and the only way to do it.

To help our situation we should have the cockpit lights turned up full bright. If a lightning strike occurs this will prevent temporary blindness. We should drop the seat as low as possible and keep our eyes inside the cockpit for the same reason. Bright lights and a low eye level help protect from a blinding flash, and they help psychologically too. The bright cockpit lights keep us from seeing out, as does the lowered seat, and this helps shut out the terror. It puts a "protective" barrier between us and that awful world out there. It's a phony barrier, to be sure, but if it helps subdue panic, who cares?

So we come closer to the actual clouds. The first thing that makes us feel something's about to happen will be a strong updraft. It probably will be smooth but powerful. The airplane wants to go up and if we push ahead to maintain altitude the airspeed will go up. We can pull power to keep the speed low, but then the engines may cool so there isn't heat enough for carburetor anti-icing. If we don't pull power then we get a nose-high attitude that's undesirable. We are probably best off by leaving the power set, holding our best rough-airspeed power, keeping the horizon on the position of pitch for level flight, and the devil with altitude. If we are using an autopilot the altitude hold should be off and the speed hold off too, if there is one. The human pilot should keep the nose on the horizon bar with the manual pitch trim. The autopilot will only be doing the heading and lateral work, which is a help and important, but it is stupid in its handling altitude–attitude–airspeed relationships.

Autopilot or hand-flown, the wings should be held as level as possible even if it takes big aileron movements and a lot of work. Nonlevel wings hurt speed control because an airplane wants to descend when it's banked; and level wings help to keep us going straight. The last thing we want to do in a thunderstorm is wander. We start in with a heading that we feel is the shortest way through and then we hold it tight to get through quickly.

The biggest updrafts occur outside the cloud. (Paul Bikel, a glider pilot, flew north along a front in the updraft for over 500 miles!) Rain hasn't started, but we don't like the feel of the updraft because we know that when it stops there's going to be a jolt. And just as we start into the cloud, our updraft stops with a jolt and the air becomes very turbulent. Side gusts, downdrafts, updrafts . . . it's wild. It's dark and as the airplane is jolted and jarred, heaved up and squashed down, the rain begins with slaps of big drops at first and then, suddenly, inundation. The rain flows over the windshield like a river. And it is noisy; most people would say it is hail. There may be soft hail and it may make a frightening noise, but it isn't big solid hail. Solid hail is a noise to end noises, it's unforgettable, tremendous, and no one ever mistakes it for heavy rain once they've heard it.

And what if it is hail? There isn't much to do. A firmer grip on emotions, but above all hold the heading! Hail is confined to a small area. Turning back will probably result in a longer time in the hail. Also, a turn will expose the side of the airplane and break out windows, and the amount of up-elevator needed for the turn may expose enough fabric, if it's a fabric-covered airplane, to tear the fabric.

I saw a DC3 once that had flown into heavy hail in Kansas, and the pilot had turned around. I looked at the airplane as it sat in the hangar, a battle-scarred wreck. The leading edges were beaten in, the landing lights broken out, and the windshield too. But most interesting was that every window on one side was broken! Fortunately it was a cargo flight without passengers. The fabric-covered elevators had numerous rips and slashes. So don't turn around in hail.

There are some interesting points about hail. Most of it in the United States occurs between the Mississippi River and the Continental Divide. April, May, and June are the worst months. The worst time of day is between 2:00 P.M. and 10:00 P.M. Cold-front and prefrontal thunderstorms are more apt to have hail than air-mass thunderstorms.

Hail forms in the storm's building stage and falls when it's in the mature stage. It is largest near the freezing level. Flight well above or well below the 0-degree point decreases the risk.

Looking again at the tropopause and thunderstorm heights will give a clue to the possibility of hail. If the radar reported storm tops push up above the trop by 5000 feet, one can expect to find at least half-inch hail.

In the black heavy rain the turbulence keeps up, but there is a feeling that it is diminishing. It's still wild and rough, but we feel more in control of the situation. There was a point, probably, in the early part when it was so rough that we doubted our ability to keep control if it got any worse. Now, in heavy rain, we feel that the out-of-control threshold has been lowered in our favor. The rain does dampen the turbulence to some degree.

SOMETHING TO BE SAID FOR RAIN

There was a time, before radar, when pilots said that if you must go through a thunderstorm pick the blackest part with the most rain.

When radar arrived this became taboo. The reason, of course, is that radar shows where a cell is by reflecting from the rain. So radar says, "There's the heavy rain, that's the storm's cell. Avoid it!" This is essentially correct. But the most severe turbulence isn't exactly in the middle of the rain. It's close by, however, and if you miss the rain area by a good margin you'll probably miss it.

Don't be sold on the idea that the center of the rain is the wildest part of the storm. The roughest place is probably in the area just ahead of the storm, near the roll clouds before you go on instruments.

Now, to go back before radar: First, everyone agreed that the best procedure was to miss the storm, but there wasn't any way of telling where the storm was. If a pilot was faced with barging in with only luck to help, he was better off picking the heavy rain area where the severe turbulence is a little less severe. Now, however, radar gives "eyes" to miss it all.

FLY!

Now back to the storm. Our job is a simple one, really: just fly the airplane. Fly attitude, keep it level and under control. We need a certain attention to the engines to make certain

the carburetor heat is sufficient and that they are not icing.
There's little navigation to do. We can glance at an omni and
try to keep on course, but we shouldn't do any excessive
squirming around with large bank angles. It isn't going to be
far enough through the roughest part to get much off course.

ELECTRICAL DISCHARGE

Lightning will flash in this darkness and some of it seems
close. The sky is lit up for moments and its brightness is scary.
Between flashes, when it's dark, we can see small flicks of fire
dance across the windshield, and a quick peek at the propellers
shows a neon-like band circling their tips. This is Saint Elmo's
fire, more formally called corona discharge. It means that the
airplane has been flying through electrical fields, absorbed
energy, and is charged. If it collects enough charge and passes
through an area in the storm that has a big charge of opposite
polarity, the energy may jump between airplane and cloud
in a type of lightning.

It goes off with a loud banging PLOP and a brilliant flash
of light. If you were looking out you'd be temporarily blinded.
But looking in or out, it will scare you.

It's over in a second and we are surprised to note that the
airplane still flies, and the wings haven't fallen off! If we
could see the damage we'd probably find a half-dollar-size hole
near the wingtip, or at the tail cone or some other small-
radius area. We might find that a radio has been damaged and
we should look with suspicion at the magnetic compass. Beyond this we are okay. Thousands of these discharges have
occurred with few serious results except as noted in the beginning of this chapter.

These are commonly called lightning strikes. They are the

only kind an airplane gets. Airplanes aren't "hit" by lightning as we visualize a person on the ground being hit. There is always an electrical discharge between airplane and cloud. A person on the ground, actually, isn't really "hit" either, but is a point where electrical potential jumps from earth to cloud through the person. Though we always think of being "struck by lightning," the person "struck" is a part of the process of discharge.

The chances of a discharge can be lessened, as we said, by staying out of the region 10 degrees to either side of freezing. The majority of discharges happen where the precipitation is rain mixed with wet snow. There are little tricks too that may help some; they certainly will not do any harm. One is to turn on propeller alcohol if available; another is to flick the mike button now and then. What these two things may do is to help carry away part of the energy that's been built up on the airplane. It's a slim possibility, but may just be enough to keep the charge below the critical point. The key thing, however, is to stay away from the freezing-temperature area. The faster the airplane, incidentally, the more likely a discharge. Also, while keeping away from the freezing level is best, it's not a guarantee. The discharge I had out of London that resulted in the big piece being torn out of the 707 occurred between two cumulus whose tops were only 11,000 feet and the temperature was 55 degrees.

The electrical charges come in two ways and they are worth talking about. One we have explained: the accumulation of a charge by flying through electrical fields. These are huge with tremendous voltages and are a part of the instability of the thunderstorm process. I always visualize them as big waterfalls of electrical energy; you fly through them and the airplane is covered by the energy and absorbs a great quantity of it.

The other kind of charge arises from flying through precipitation. This, crudely, is a friction process, like the crackling one hears when combing one's hair with a plastic comb on a dry day. This type of charging doesn't affect things nearly as much as the field type.

It's easy to see that when flying through a thunderstorm we get the field-type charge, and if we are flying in the wet-snow area we also get the friction type in its worst form; so we are getting the maximum. Getting away from wet snow deducts part of the charge—unfortunately the smaller part.

STATIC AND RADIO

All this charging causes static, which knocks out radio reception on certain frequencies. Fortunately, it rarely affects reception in the VHF ranges. It offends most seriously in HF frequencies, and in the old 200- to 400-kilocycle radio-range frequencies.

An electrical charge looks for the easiest way off the airplane. This will be any small area like a copper antenna wire. Of course when the charge bleeds off an antenna it makes a terrible noise in the radio. To prevent this a large wire with a polyethylene coating is used, the entire thing being about the diameter of a pencil. Also, the insulators and attachment hardware are large and smooth so that no wire end sticks out in the air. The big wire and smooth insulators discourage discharge.

The other part of antistatic hardware is the little wicks we see sticking out from wing and stabilizer tips, and sometimes from the rudder and tail cone also. These are designed as a place where the charge can bleed off quietly and easily.

All this helps, but only for the "friction" type charging; it doesn't impress the electrical-field type at all because it is so big the wicks cannot bleed off; nor can antennas contain it. If they could, we'd never have a big bang discharge.

This was a major problem in the radio-range and HF days. It is much, much reduced now because of VHF, but electrical fields are so strong even VHF will start to howl and squeal in thunderstorm conditions at times. It generally doesn't last long, but when one can hear it it's likely a discharge may happen! For long-range airplanes needing HF for communication and Loran for navigation, precipitation-static discharge is a problem.

The ADF will be knocked out by precipitation static, and it's important to keep in mind that ADF bearings will be very unreliable. Following an ADF indication could easily take one off course and perhaps set up a dangerous situation, particularly in mountainous areas. Look at all ADF bearings with much suspicion when in static.

A THEORY

Before the coming of radar we developed a theory, born of desperation, that the ADF would show one where thunderstorms were. The idea was to set the ADF so it was not turned to any station and preferably on a low frequency— around 200 kilocycles. Then we studied the ADF needle movements for any steadiness in one direction. This was to indicate, roughly, where storm centers were. It had to be done before any precipitation was entered and static knocked out the ADF, because once precip-static starts, the wandering ADF needles don't mean a thing.

This theory of pointing thunderstorms may have had some basis. Personally, I never had much luck with it, but others claim they have.

There is a new thunderstorm avoidance system, developed

within the last few years, that zeros in on lightning and shows its direction and, vaguely, its distance from the aircraft. Since thunderstorms have lightning—if they don't they aren't thunderstorms—the idea has merit and reports to date show promise. Like all devices, it requires experience and learning to use well.

This system has the advantage of not needing an antenna, like radar's moving dish, so it's easily installed in single-engine airplanes that don't have the nose for radar, or that must have the antenna mounted in an external "bomb" on the wing, which adds drag.

This new thunderstorm avoidance concept is worth keeping tabs on.

THE NOISE IS ANNOYING

While we are tossing around in the dark, wet, and rough inside of a storm, we can use radio noise in a couple of ways.

If we have HF frequencies, or are using radio-range frequencies, the noise will be so continuous and loud that nothing else will be heard. All the radio is doing then is aggravating our condition of apprehension. The best thing to do is to turn it off. It's surprising how calm it is, even in the turbulence, to suddenly have that infernal noise go away.

If we are using VHF, we leave it on. The occasional squeaking noise we hear may tell us a discharge is close. While at this point we cannot do much about it, it will at least have us prepared and not so startled when the bang and flash occur.

ALMOST THROUGH THE STORM

As we continue the rain slackens, the turbulence quiets down and patches of lightness bring encouragement. We may

fly out of the storm quite quickly and dramatically find ourselves in brilliant sunshine. While we breathe a sigh of relief
it's a good idea to remain fastened down and alert until well
away from the area.

We have traversed this storm from front to back. The worst
part was first because the roughest part is the front, or leading
part, of a storm.

Had we been coming the other way the sequence would
be reversed. The rain would begin, and turbulence start; it
would become darker and darker and the rain and turbulence
heavier. Toward the front side, light spots would appear, but
it is then that we'd get the real rough stuff, that wild updraft
and turbulence even after we have broken out of the clouds.
Flying back to front we want to be well through and ahead of
the front before we relax.

Now suppose we are flying a jet and hit all this up high, at
25,000 or 30,000; what's up there? Lots of snow, for one
thing, and turbulence, probably not as rough as down lower,
but still very rough. There is more chance for hail. The massive amounts of snow, some of it in the form of very large
flakes, may cause icing, especially inside warmer intakes where
the snow is melted only to refreeze later in a colder area.

While doing thunderstorm research I barged into one at
35,000 feet in a B17. It was impressively rough, but most impressive was the size of the precipitation: the snow looked
like big snowballs being thrown at us. This stuff got into the
intakes, clogged them up, and by restricting the airflow caused
supercharger ducts to squeeze in. The result was four very
weak and helpless engines and a big, not-so-good glider! Fortunately the high altitude made it possible to glide away from
the storm, get lower, and reorganize things so that we could
limp to an airport.

Up high or down low a thunderstorm is potent. Even over

the top of one, turbulence will extend upward into the clear air a few thousand feet.

WARM-FRONT THUNDERSTORMS

When flying warm fronts we will be on instruments most of the time and it will be difficult to see individual storms. Occasionally the airplane will break out between layers and then it may be possible to see the CBs towering up from the bottom layer into the higher layer.

Warm-front thunderstorms do not seem to contain as much violence as cold-front storms. They will be rough, no doubt, and it will rain hard and lightning will flash, but the real sock—bang stuff isn't as bad. Light rain may fall, and occasional crashes of lightning are heard in the radio. If we approach a cell, the rain will increase and the radio noise will become louder and more overpowering.

Because we cannot tell without radar where the storms are in a warm front we should fly prepared to get in one any time. And while they may seem a little less violent, they are still thunderstorms!

LOW DOWN

The best rule is to fly low. Keep down, near the minimum instrument altitude. Since we are dealing with turbulence, it's best not to fly *too* low, and I personally prefer 2000 feet above minimum instrument altitude. This is dependent on terrain. If it's all flat I'd feel secure 2000 feet above the ground, but in mountainous areas I'd want more.

This low flying in warm-front thunderstorms is due to the fact that warm-front thunderstorms generally have high bases. The air has to crawl up the warm front to have enough lifting to set off the thunderstorm.

If we are flying a jet we don't fly at 2000 or 4000. With its high-altitude capability a jet will sometimes top an entire warm front—over the ocean most certainly. But if a warm front with thunderstorms cannot be topped, then it's best to come down to an altitude where the airplane is flying well and to use radar to go through. The best altitude will depend on the jet's performance; generally speaking, about 33,000 feet is good.

THUNDERSTORMS AT LANDING

Sometimes we are faced with the problem of our destination being covered by a thunderstorm, or having them so close that they affect our landing. What do we do?

First, we must approach the area and dodge any cells near the airport. We can do this by VFR procedures, if we can see, or by radar. The complication, of course, is traffic control. If we wander around dodging storms we have to keep ATC advised and get their prior okay. It's only logical that two airplanes from opposite directions might be headed for the same good area. A deviation also might put us on another airway.

ATC will sometimes help lead one through a thunderstorm area. This can be a mixed blessing. The bad part is that air-traffic radar does not clearly show thunderstorms. The equipment is designed to show airplanes and not weather. The storms show in a vague way on the scopes and some controllers feel they can use this limited picture to lead you through. But they cannot do so adequately unless there is a special weather radar nearby.

The best plan is to ask the controller, when he begins steering you through a thunderstorm area, if he has weather-radar capability. If he doesn't, you are probably better off doing it yourself. The FAA is trying to cure this problem and we wish them speedy good luck.

DON'T RACE THUNDERSTORMS

A real hazard is a pilot trying to beat a thunderstorm to the airport. There's nothing wrong with trying to get there first, but there are a couple of possible pitfalls. One is diving and allowing the airspeed to get too high, and then suddenly flying into the rough area just ahead of the storm at this high speed. Another is racing to get to the field and then landing as the storm arrives, or even when it's 5 or 10 miles from the airport. The wind shifts abruptly with gusty force, and a cross tailwind suddenly makes landing difficult, perhaps impossible. This, again, is where severe shear occurs.

With the wind shift comes heavy rain that obstructs all vision, and one may suddenly go completely on instruments— zero-zero—anywhere along in the approach. At 100 feet, for example, the pilot will suddenly be unable to see the ground. Even if the rain arrives just as he touches down, the visibility will be so poor that it will take luck to make a straight runout and stay on the runway. It's a terrific shock to suddenly learn how bad visibility is in heavy rain. Braking will be poor, possibly nil on the wet runway. The airplane can aquaplane and not stop. With a strong crosswind it can also slide off the runway sideways.

HOW TO TURN AROUND

Now, an important point. Let's say we are racing a line of thunderstorms to an airport. The line arrives just as we do.

We decided to pull up and abort because landing is impossible. Now, which way shall we turn, right or left? In the northern hemisphere it's best to turn right. Why?

The wind has probably changed to NW. Making a left turn, we will have a momentary drop in airspeed, bringing us closer to stall; the sudden new side component of wind will also tend to overbank the airplane; an intense downdraft will be present as the cold air arrives. It all spells an airspeed loss that's difficult to regain. There have been accidents that were caused by loss of control when turning away from a thunderstorm.

Of course, if one is still in the southerly wind part of the approaching storm he'd best turn left. The point we are trying to make is that if you get in close to a storm, particularly where the rain has started and the wind shifted, and then decide to turn around and get out, it is best to make a right turn.

There isn't always time or circumstances to do the correct thing, but right or left, remember the air will be gusty and shifty and it's very important to maintain enough airspeed above stall to take care of downdrafts and wind shears. If power is needed to accelerate the airplane, this isn't a time to be bashful about the amount. Pour it on, lots of it. It takes an airplane longer than you think to gain airspeed by power alone when pulling out of a low-airspeed condition.

AFTER WE'VE TURNED AROUND

When we have gotten turned around and away from the storm, what then? If it's an air-mass storm we may want to hold out in a clear area until it leaves our airport, and then go in and land. If it's a front, we'll have to pass through it before we can get to our airport. This case calls for a retreat to some still-clear airport where we can land and wait until the front has passed. Then we fire up and complete the flight calmly and pleasantly.

In summary, the first thing in any airplane at any time is to stay out of thunderstorms. If you must fly through fronts or heavy cloud cover that contains hidden thunderstorms, don't do it without radar. If the radar fails while you're in the middle of the mess or for some other reason you get caught, fly the best speed, fly attitude, and hold a heading: Remember power settings and stabilizer settings so that if excessive dives, climbs, or upsets occur, you'll know where things ought to be reset! Don't fly very low, or very high! And most important, keep emotion under control and fly smoothly, scientifically, and coolly.

13 Ice and Flying It

WHAT'S IT LIKE when an airplane picks up ice?

It begins by forming on little corners of the windshield, and on the wing we see a polished look on the leading edge if it's clear ice, or a fine line of white, if it's rime. If the windshield isn't heated, a smear of ice will cover it.

Ice will begin to accumulate on pieces of the cowling, masts that stick out, and other un-deiced places.

An aware pilot knows before he gets into this what his indicated airspeed is, and also the RPM and Manifold Pressure of the engine; as ice forms he will keep a constant check on them to catch any change.

This early stage is the time to do something about getting out of the ice. Get a different altitude. Go down if there is an above-freezing temperature below. Get on top if you know where the tops are. Be careful about climbing, however, if you don't know what's up there. If it's a warm-front condition you can climb into worse conditions, and climbing along the slope of a warm front is really asking for a load of ice. As a rough rule, we climb in a cold front and descend in a warm front, but this is variable and each weather condition should be analyzed and studied by the pilot before he decides what to do.

Unfortunately we cannot make hard-and-fast rules about

each weather condition. We know things in a general sense, but we have to learn by experience.

Fortunately, fronts take up a relatively small portion of our weather time. The lesser weather, such as air-mass conditions, makes up most of it. We can come closer to rules for flying in these conditions, and they are covered in various places in this book.

But let's imagine we got into ice and didn't do anything about it, but instead just sat there and let it grow.

We would begin to see an airspeed drop. This is always a little alarming. A couple of knots doesn't impress us much. Five knots begins to, and when ten knots have gone we get edgy.

The thing to do is pour on more power. We want to keep the airplane at as low an angle of attack as possible. We don't want to get in a tail-draggy situation with higher drag and ice forming back under the wing.

About the time our airspeed loss is bothering us, we develop a bad vibration. It is a hunk of ice breaking loose from the propeller. A little more propeller alcohol is needed if we have it, or more heat.

Now another vibration begins. The unheated windshield gets a thicker coating and we can't see ahead at all. On the corner of the windshield a big hunk of ice has built straight ahead into the airstream, and it's four to six inches thick!

The vibration gets worse and now a howling starts.

The howling is probably an antenna mast vibrating from ice disturbing the airflow around it.

The wings have a good coating, and if we have rubber pulsating boots we turn them on. The ice breaks off in chunks, but leaves pieces that sit in the airstream and go up and down with the boots. We turn them off.

The engine cowls now have a large coating of ice on their

leading edges. The scoops have ice around their openings. The propellers periodically heave off hunks of ice and the vibration increases. The antenna mast howls and shakes.

The wing has more ice and we try the boots again. Hunks break off, but some hunks don't. There's a messy collection of pieces of ice behind the boots.

The airspeed has dropped more and we apply more power, resetting the carburetor heat as we do. The throttle is open an alarming amount.

The ice on the engine cowls is almost out to the props. The vibrating and howling from masts is eerie, and then it suddenly stops. So does one of our radios. The mast has carried away.

The wing boots are only doing a part job, and bigger hunks of ice go up and down with the boots. The airspeed is decreasing more and the throttle is wide open. The only way we can maintain speed is by losing altitude. The only way out of this mess is down and we hope there are above-freezing temperatures down there before we get to the ground.

The situation is pretty desperate, but unnecessary if we had done something about getting out of the ice when that little bit first formed on the windshield corner.

Ice affects the flying qualities and characteristics of an airplane. The most serious thing it does is destroy smooth flow and make a different airplane of the one we know. The weight of ice is of secondary importance.

Ice affects the wing section. It affects the propeller in the same fashion. It collects on things sticking out that create parasite drag like antenna masts, wires, and cowlings, and makes them drag more.

Ice will also cause an object like an antenna mast to vibrate and howl severely and finally to break off, or half off, which may be worse. The first experience I ever had with ice was in a Pitcairn Mailwing biplane. Cockily, with a newly learned ability to fly instruments, I barged into a large midwinter CU. In a few moments I had an appreciable load of ice. It collected on the wires which held the wings on. The wires started to vibrate wildly up and down at least a foot each way. I expected them to break at any instant. I didn't have a parachute. I quickly flew out of the CU, the ice shook off, and the wires didn't break. I had become much older and wiser.

The next serious ice experience involved flying a Constellation across the North Atlantic. At 18,000 feet I encountered light ice, which covered a long antenna that went from a mast up front to a fin at the back. The antenna vibrated and finally broke near the fin. The long wire, with a big anti-static insulator on the end, whipped around beating the fuslage. I tried different airspeeds in the hope that the wire would "fly" out straight and stop beating the fuselage. Nothing helped and finally the insulator gave a sound whack to a window and broke the outer pane; fortunately there were two. But this meant I had to reduce the cabin pressure and descend. I descended into a worse icing condition. It was a moment of relief when we finally landed at Gander, Newfoundland.

The point is that often the gadgets on the airplane are the worst offenders, and if one expects to fly a lot of weather he should try to make the airplane as clean as possible, with a minimum amount of stuff sticking out.

Flying a 747 I've had little ice and none of the howling, vibrating kind from gadgets that collect ice. The jets are "clean" airplanes and the "cleanness" reflects in their resistance to icing.

It is important to realize that deicing equipment does not allow one to fly indefinitely in ice. It cannot do this job! It will help and give one time to work his way out of icing situa-

A radio mast, about twelve inches high, loaded with drag-causing ice. Things like this can't be deiced. The more of these "ice catchers" on an airplane, the less time one can stay in ice—deicer-equipped or not. (NOAA PHOTO)

tions, but it will not allow one to sit there all day long. There are a lot of reasons why it isn't good enough, and one of the main ones is that deicing equipment doesn't cover the entire airplane. Another is that it doesn't clean off all ice completely.

So we have an important first rule: When ice is encountered, immediately start working to get out of it. Generally this means a different altitude, after a request to ATC. Unless the condition is freezing rain, it rarely requires fast action and certainly never panic action; but it does call for positive action.

Now let's talk about ice in detail. First, there are two kinds, clear and rime. They are what they sound like. Clear ice is a smooth, hugging type that is tough. Rime is crystal-like and pretty. It often forms in weird shapes and sometimes sticks out as long cones into the airstream ahead as it forms in windshield corners or other places where airflow patterns change. Rime ice breaks off fairly easily with deicer boots that pump up and down.

Clear ice doesn't, and one description of the difference between rime and clear is that clear ice cannot be removed by deicers and rime can. This is an oversimplification, however.

Ice comes in three classes: light, moderate, and severe. These are difficult classifications, which depend on the pilot's judgment; one person may think a certain degree of icing is light, while another calls it severe.

Generally speaking, the categories refer to the rate at which ice forms. In flying through freezing rain, ice accumulates very fast, and is called severe. Since it's rain, it has to be falling from an area warm enough to have water. The cue in freezing rain is to go up. When it first forms we might also go back, or land. At any rate, we must do something fast because it forms fast!

I remember a DC2, flown by a friend, approaching Chicago's Midway. At McCool he ran into freezing rain. The field was so close he decided to make a run for it. He landed with almost full power dragging him in. There was so much ice on the airplane that it was necessary to chip it off the fuselage to get the door open! A few more minutes and he would have spun in. He said he'd never try that again. This was during the time, before we knew better, when we thought a DC2 could fly through *any* weather. We learned differently.

Light ice would be found in a thin stratus cloud in very cold air that doesn't have much moisture. Light ice can, of course, become moderate if we sit in it long enough.

The speed with which ice forms is the thing that interests the pilot, because it affects what he's going to do about getting out of it. If it's light he can move slower, talk it over with ATC, and have time to work. If it's severe, however, he might be in an awful hurry, and just tell ATC what he's doing and that it's an emergency.

Before we go any further let's talk about deicers. What do we deice?

Pitot heads of course, carburetors, or in the case of a jet engine, the engine inlet cowl, guide vanes, etc. This is obviously important because if the engine doesn't run, we don't fly.

Carburetor heat can be tricky. First, one should have a carburetor-air-temperature gauge, and then know exactly where the temperature bulb is located in the carburetor system. The reason is this: If the bulb is located in the coldest part of the carburetor, one only needs to pull on heat until the temperature gauge reads something over freezing, like 35 degrees F. If the bulb is located in the scoop, it may be necessary to carry as much as 85 degrees indicated to be certain that the coldest part, farther downstream, is above freezing. There is about a 30- to 40-degree temperature drop from the air scoop to the coldest part of the carburetor.

This can lead to trouble if there is not sufficient heat. A pilot might be flying in a condition of cold dry snow, for example. Generally this condition doesn't cause carburetor ice, because the cold ice crystals zip right through the engine and don't get up to a slushy, freezing stage. But if, in this very cold air, a person put on heat as a precaution until a bulb in the air scoop read just above freezing, he'd raise the temperature enough so that the ice crystals would get soft and then refreeze in the colder part of the carburetor system and cause trouble.

It's therefore important to learn where the carburetor-temperature bulb is located and how much temperature drop there is through your system, and then to use carburetor heat accordingly, always having it above freezing in the coldest part of the carburetor if conditions require heat.

Under most conditions it's obvious if heat is needed. It's used, for instance, during wet snow, in rain, in cloud with a near-freezing temperature, even in clear air with high moisture content if the carburetor is a sensitive one.

A sunny summer day may produce carburetor ice because warm air can hold a lot of moisture; so if it's warm and humid, carburetor ice is a real possibility, especially at low power, such as when idling during the glide on an approach to landing.

But sometimes it is difficult to know whether heat is needed or not. A close watch on engine performance will tell.

If the airplane has a fixed pitch propeller, watch the RPM carefully to see if it's decreasing. (To do this a pilot has to know the RPM he has set and also be certain he isn't climbing, because even a small climb angle will decrease the RPM.)

If we get an RPM drop, it's a good indication of ice. Pull on carburetor heat. The RPM will drop further because the mixture has been upset with the addition of heat. Leave the heat on for a minute and then take it all off. The RPM will bounce back to its original value if the ice has been melted out. This tells, of course, that ice was the problem.

Naturally, we don't want to operate with this off–on procedure; so now, knowing there is ice, we pull on enough heat to get our carburetor temperature above freezing in the coldest part of the carburetor.

The RPM will drop when we do this, because the mixture has become rich. The hot air we are now using is less dense, so there's effectively less air in the mixture. We lean the engine with the mixture control, and when we do so, most of the RPM loss should come back.

If we haven't a carburetor-air-temperature indicator, then we put on heat by guesswork, lean out, and keep a close eye on the RPM. If it holds, we've got enough heat. If another drop occurs, we should pull on full heat to clean out the ice, and then apply heat again, except this time a little more.

We don't want too much heat; we ought to try and use only just enough. Generally, if too much heat is used, leaning the mixture will not bring back the original RPM. If excessive heat is used, the engine may run rough and lose power.

If our propeller is the constant-speed type, we'll never notice any RPM change because as the engine loses power the propeller pitch changes—gets flatter—and keeps the same RPM.

The way we tell, then, is by watching the manifold-pressure gauge for a drop. That will tell us there is ice. Then we go through the same procedure as mentioned before except that we use the manifold-pressure gauge instead of the tachometer to judge power loss or gain.

If we change altitude we'll have to reset the heat–mixture relationship once leveled off at the new altitude. Most engines use rich mixture for climb, too. With high-powered outputs, as at takeoff, carburetor heat isn't needed, and if one used it on takeoff there would be a big power loss, and even a chance of engine damage. So we clean out the ice before takeoff, take off without heat, and once in the air watch closely for any signs of engine icing.

During climb most engines do not use heat, but under severe conditions it might be needed. Any deviation from normal climb performance, on instruments, is a hint of carburetor ice.

We hear a lot about nonicing carburetors. These are injection types and don't have a carburetor venturi with its temperature-lowering action. These nonicing engines don't ice very easily—they seldom do at all, in fact—but icing can nevertheless occur in them because it is still necessary to bring in air from the outside for the engine to breathe. If this air has the proper moisture content and temperature, freezing may occur somewhere in the induction system.

Clever duct design has reduced this hazard to a minimum, but icing can and does occur and a pilot still has to keep the possibility in mind. Because advertisements say "Nonicing" we should not go along innocently believing it 100 percent true!

In the awful condition where there isn't enough heat to clear out the ice, a desperate trick might help. The trick is to try to make the engine backfire in the hope that the backfire will clear out the ice. Generally this can be done by leaning the mixture until the engine runs rough and backfires. If it's successful, the engine will come back in with a great roar as you enrich the mixture again. I've done it twice under desperate conditions, once in a DC2 and the other time in a little Culver Cadet. Both times it worked.

Jet engines are a much simpler matter. There isn't any

carburetor and so there isn't any carburetor ice. But the air inlets, cowling, and guide vanes can collect ice just as a wing does, and sometimes even when the wing doesn't, because there is a temperature drop in that big, venturi-like cowling that can cause ice occasionally, even in clear air if the air is very moist and the temperature just right.

Ice affects jet engines seriously as the inlet airflow is disturbed and cut down.

Fortunately the engines are equipped with hot-air passages within the cowls, guide vanes, and so on. The pilot simply flips a switch and hot air from the engine is routed through the passages to keep them warm. The only effect on operation may be a slight power reduction from air loss in the engine. The loss is almost insignificant.

Jet inlets can ice up on the ground during a long ground hold, waiting for takeoff, during high-moisture, lowish-temperature conditions. The jet inlet causes an air temperature drop and ice may form on its walls in the above conditions. The airflow into a turbine engine is very critical; a relatively small amount of ice can disrupt it to the point that when we're finally cleared for takeoff and advance the throttles, we may get irregular engine action or a compressor stall.

So while holding on the ground it's wise to use cowl heat periodically and prevent ice accumulation. There doesn't have to be visible moisture such as rain or snow, but simply high humidity—cold fog would be suspect too.

The main point is to be certain that heat is used whenever there is a possibility of icing. All engines, piston or jet, get their deicing power from heat generated by the engine, so it's obvious that if the engine isn't running there isn't any heat. Don't let it lose power.

We now have pitot heat on to keep the necessary instruments working, and engine heat; next in importance is the propeller. Jets don't have this problem, of course, and that's a very nice part of the jet world.

THE PROPELLER IS IMPORTANT

Propellers ice, and if they do their efficiency drops quickly and then the engine we've kept running is just slinging an icy club around beating up the air, but not doing much pulling. Ice can also unbalance a propeller and make the engine run rough enough to shake your teeth.

There are two kinds of propeller deicers: alcohol that's sprayed on the leading edge of the propeller to melt ice, and heat in the leading edges.

There are chemicals available which can be spread on the leading edge of a propeller to help keep ice off. These same products, when rubbed into the surface of the wing deicer boots, make the ice come off more easily, since pieces of ice don't tend to stick to the boot.

The preparation doesn't have a very long life and is washed away when the aircraft flies through rain.

There are two ways to use deicing equipment. One is de-icing and the other is anti-icing. In the first case one waits until ice has formed and then turns on alcohol or heat to get rid of the ice. Unfortunately the ice never comes off evenly and when it comes off one propeller blade and not the other the unbalance makes a terrific vibration. Also, on a multi-engine airplane the ice slinging off often hits the fuselage with a loud whack that's disturbing to one's nerves. It doesn't hurt the airplane particularly, but any old-timer DC3 will show dents on the side of the fuselage next to the propellers where hunks of ice have beaten on it.

Sometimes, too, the ice is tough and will not all clean off properly.

The better method is to use the equipment as an anti-icer. This simply means to get alcohol or heat on before entering icing conditions so as to keep the ice from ever forming.

If there is ice on a propeller and the deicing method used is having trouble getting it off, try running the RPM up and down in surges to give extra, and irregular, centrifugal force, to help sling the stuff off.

Propeller ice will often form before visible wing ice, and if one is flying in cloud and an airspeed loss occurs without wing ice, it may be because of propeller ice. I saw this when doing research work with a B17. We had a stroboscope arrangement to look through. It "stopped" the propeller visually while it was spinning at its normal RPM, and you could look at the prop as though it were standing still. A B17 was convenient for this because you could see the props closely from the navigator's station in the nose. It was amazing to me to see ice on the leading edge of the blades when we didn't have a bit anywhere else on the airplane. And this happened quite often.

So don't be bashful about getting the propeller anti-icing gear in action in advance or under any suspicious condition. The propellers are most important and if I could have either wing deicers or propeller deicers I'd take the propellers. If they are doing their work efficiently you can pull a lot of ice-covered airplane around the sky.

WING DEICERS

We have two major sorts of wing deicers, pulsating boots and hot wings. The boots are always a deicer.

Many times boots are used improperly, and if used improperly they can cause more trouble than no boots at all. The boots pulsate up and down and, in theory, break off the ice. They don't break it off as it forms. The boot has to wait until there is a coating of ice before it can do an efficient job. It is therefore advisable to allow ice to build up at least one-quarter of an inch thick before turning on the boots. They then expand and break the ice, and the windstream blows it off and away. When they have done this job, it's time to turn them off and wait until another coating of ice has formed and then turn them on and break it off again. Some late model deicer boots have an automatic off–on scheduling device so that the pilot doesn't have to worry about it, but I've never thought much of this gadgetry.

DO BOOTS DO IT?

Actually, I've never thought much of deicer boots. In thousands of hours of weather flying I've never seen a situation where I couldn't have made it without boots and most times I've turned them on have been for amusement.

A good example of this occurred in the B17. The boots, during those difficult times of World War II, were made of something less than the best rubber and we continually had problems with them developing tears. Each time we had to replace the boots it was a long maintenance job. Finally, we had a bad tear and couldn't find replacements. I told the crew chief to take them all off and we finished 18 months of weather flying, a lot of it in ice, without boots on the airplane! I was particularly fussy, however, about keeping the propellers clean and the engines running.

Of course the B17 had a thick, relatively high-lift wing. The modern laminar-flow wings are more sensitive to flow disruption, and boots may be needed. However, care must be taken to be certain the ice is broken off clean by proper boot use technique. A dirty leading edge with ragged pieces of ice sticking to it is often worse than a smooth coating of ice. Boots, at best, are not the most desirable way to deice a wing.

At the risk of boring the reader I repeat that the worst ice offenders on the airplane are the aerials and gadgets and places where boots cannot be.

I've seen cases where the airplane had a smooth coating of ice when the boots were turned on. Then it broke up in chunks with pieces left on the airplane, even big hunks that stuck on the leading edge and went up and down with the boot, but never blew away. I've seen the airspeed go off ten knots after the boots were turned on and the ice cracked.

Boots raise stalling speed to some degree and must not be on when landing or taking off.

HOT WINGS

Hot-wing deicers are something different. They heat the leading edge. They can be used as deicers or anti-icers, but are better used as anti-icers.

When used as deicers, that is, turning them on after ice has formed, they melt the ice; but when it melts, the water often runs back on the wing and refreezes and forms a spoiler on the top of the wing that can materially affect airplane performance. (If this happens, try lowering the flaps a little to get back some lift. Watch flap extension speed restrictions.)

Hot wings are more often, and properly, used as anti-icers. They are turned on and the wing heated before getting ice. The hot wing never lets ice form.

YOU HAVE TO SEE

An important item that needs deicing, or anti-icing, is the windshield. If we make a low approach through icing conditions and get the windshield covered with ice, we cannot see to land.

The ways to cure this are:

1. Hot windshield.
2. Alcohol to squirt over the windshield.
3. A window one can open to see out.

Historically we have handled the windshield-ice problem in the reverse order. Back in DC2 and early DC3 days we all carried a putty knife in our flight kit. It was used when ice covered the windshield. We'd open the side window, reach out into the icy blast around to the windshield and scrape off enough ice to give a small clear area to peek through while landing.

Then we got alcohol. This would squirt across the windshield and as it melted the ice the windshield wiper would carry it away or knock it off. It was smeary and partially effective. We still carried putty knives.

An occasional heavy coating of ice wouldn't yield to either putty knife or alcohol, and there have been cases where the pilot bashed out the windshield with a fire extinguisher to have a place to look out!

Now we have heated windshields that seem to do the job well. All the trying days of fire extinguishers, putty knives, and alcohol are taken care of by flipping a switch.

However one does it, keeping the windshield clear is a very important part of ice flying. If the methods available are marginally adequate, then a window that can be opened for viewing is a must! I've always disliked flying any airplane that didn't have a window I could open and at least get enough of a look through to land. A covering of oil from a bad leak can make a windshield no more transparent than a solid wall!

These are the protective devices against ice. They don't cover the entire airplane. Even if propellers and wings are kept clean, ice will build up in other places, causing drag that eventually will be very serious; which gets us back to the original statement that the first rule of flying ice is to do something about it as soon as it occurs.

HOW WE FLY ICE

Ice flying begins before we ever leave the ground. A number of things need checking.

First and most important, is there any frost or ice coating on the wings? If so, we have to get it off before takeoff. The rough surface of the ice can ruin the airflow over the wing, so that the takeoff is long or even impossible; airplanes have run the full length of a runway, through the fence, and out into the boondocks because the frost on the wings would not let them fly!

The propellers should be clean. On damp, misty, coldish days it's wise to use anti-icing on the propellers right from the start of takeoff. There is some evidence that one can get propeller ice in clear, but very humid, cold air.

There's indication that under these conditions a jet engine inlet may ice, and so engine heat may be wise with a jet engine also under these conditions. Coldish can be anything from roughly 25 degrees F to 45 degrees F.

The airplane's controls should not have any ice obstructing their movement, and the landing gear should be clean, especially if it's retractable.

Check the pitot head and static sources closely to be certain they are not blocked by ice or frozen slush thrown up by the wheels during a previous landing, or by precipitation which stuck on the airplane while it was standing.

If it's very cold be certain there has been heat enough, and applied long enough, to warm up the instruments. A cold gyro may be slow in coming to speed and its action consequently sluggish—which would be bad if one took off and went on instruments quickly.

The windshield may fog up during takeoff, and we'd best be prepared to defog it with whatever means are available.

After an engine is started it's important to warm it up thoroughly, so that when we take off the engine will put out, and continue to put out, its maximum power.

Carburetor heat may be necessary during warm-up and taxiing, but the heat should be off for takeoff; it should be off for taxiing, if possible, but there should be a short period of hot air on just before takeoff to make certain that all ice is removed.

ICE FLYING STARTS ON THE GROUND

Taxiing can be quite an interesting experience on icy and frozen snowy runways.

First, if there's snow and it has been plowed, it is important to be certain wingtips, and the tail when turning, will clear the snow banks—especially if we are in a low-wing airplane.

The runway surface may be ice-covered and slippery, and a combination of slick surface and strong wind trying to weathercock the airplane makes taxiing an adventure.

First taxi slowly, really pussyfoot along. The most important point is to use the brakes carefully! Use them in little bursts. Don't lock the brakes. As soon as they are held on, the wheel stops and slides, ice-skate fashion, over the surface. Tapping the brake on and off at short intervals will give braking for an instant each time the brake goes on, just before the wheel stops and becomes a sliding ice skate.

I operated my B17 in the Aleutian Islands of Alaska. The weather there often causes sheet ice to cover the runways and taxiways. The winds blow hard, the taxiways are narrow, and a B17 with its big fin wants to swing around like a weather vane. I had excellent luck getting around by using the quick on-and-off braking method.

Sometimes on slick runways it's impossible to get the brakes to hold while running up the engine, so one might have to run them up while sliding. This takes thought and planning and careful observation of where one's sliding.

The start of takeoff can be swishy as a crosswind tries to slide the airplane around before there's enough airspeed to make the rudder effective for steering. It takes care, and the winter isn't any time to do things in a rush-rush fashion.

Turning onto a slick runway at high speed for a running takeoff is a very poor practice, because the nose wheel will not have any traction and will only skid as you use it to straighten out the airplane and aim it down the runway. The nose wheel may turn, but the airplane will not, and everything will slide sideways out of control, headed for the boundary lights!

WHERE WE FIND ICE

Let's look at the places we find ice and how to get out of them.

As always we are back to our first look at the general weather picture to see where the fronts are, because in the fronts and low centers we find the most ice and the ice most difficult to avoid.

Ice can be found in large amounts out of fronts too, but it's easy to avoid if a pilot understands the weather picture. Ignorance can find a pilot desperately in trouble when he need not be.

The classic case of this is the Allegheny Mountains after a low and its associated fronts have gone off the East Coast out to sea.

Although the fronts have gone we find the country from Harrisburg, Pennsylvania, to Columbus, Ohio, cloud-covered. The mountain weather-reporting stations are grim: Low ceilings, snow falling, and visibility nil. A pilot flying low, trying to stay VFR, would have a desperate time. It would be impossible to cross the ridges as they would be in cloud.

A pilot flying a few thousand feet higher, on instruments, would find himself getting ice at an alarming rate.

But a clever pilot would know that all this was an air-mass condition with reachable tops above, and he'd be sitting in sunshine, on top of a seemingly endless blanket of white, relaxed and comfortable.

Such air-mass stratocumulus decks are common in winter after fronts pass. The new cold air is unstable and builds a cloud deck. On the ground, after a front has passed, we are often under clear skies. But cumulus start to build. At first there are pretty white scattered CU; they become broken and dark and finally turn into a gray overcast sky that spits snow in blustery cold winds. This is the real cold air mass moving in.

That cloud layer has ice in it. The only decent place to be is on top. The tops will vary in height depending on a number of things.

Tops are higher in mountainous areas or to the lee of large bodies of water, such as the Great Lakes, where the air picks up moisture. The tops will be higher closer to the frontal system, though not directly behind it, because there is often that clear area where the new unstable air hasn't had enough time to get in and start its action. This clear area isn't very wide. It often fools people into thinking everything is wonderful ahead, that the front has passed and now it's clear. These hopes fade as clouds begin to form and bases get lower and tops higher 100 miles or less along the way.

As we get farther and farther from the weather system the tops are lower and the bases higher, and the showers less frequent until they die out altogether. This happens because the air has been in the area long enough to be modified and its instability is reduced as the air temperature and ground temperature become more nearly the same.

TEMPERATURE AGAIN

This temperature difference between ground and air is the reason why we often see beautiful, clear, cold, winter nights and, as we look at the sparkling stars, decide tomorrow will be a lovely day. Tomorrow turns into an overcast, cold, gray day. Why? Simply because when the sun came up it warmed the ground; maybe it didn't feel warm, but it was warmer than the air, and this started the air rising and triggered the same process that makes cumulus clouds in summer.

WHERE ARE THE TOPS?

If we are going to fly in the area of the heavy deck with its ice and snow, we should try to learn what the tops are before we take off and start climbing up to find them. Ask the FSS or the Weather Bureau. If they don't know, ask an airline pilot who has just come in, or call the operations department of an airline and ask them to ask a pilot who's taken off or landed. The airline pilot is happy to help out. And when you get up there don't keep it a secret, call the nearest Flight Watch or FSS and tell them what ice you found climbing up through and what the tops are.

If we are climbing to top a deck it's best to climb quickly—not at a nose-high staggering angle so that ice will form back under the wing and hurt, but at high airspeed and with plenty of power.

Stratocu clouds have the most ice near the tops, so don't struggle along clipping through the tops; get up and out of the clouds.

Occasionally there are bits of trickery in getting on top. Out of Pittsburgh, Pennsylvania, headed east, the highest tops and toughest ice are located in the mountains east of Pittsburgh, over Laurel and Chestnut Ridges. Taking off and then climbing toward the east means one climbs up through the thickest clouds. To the west of Pittsburgh, of course, the land is lower. Back in DC2 and DC3 days we often got a clearance to take off and climb toward the west. The tops would always be lower that way. Once well on top we'd turn back east for New York. It was a way to get up through the minimum amount of cloud. The tops to the west would often be 7000 feet while the tops over the ridges might be 12,000 feet.

Flying in winter over the mountains of the Far West can be difficult, with a combination of moist Pacific air being lifted up the mountains of the coastal ranges, Cascades and Sierras. It can be a problem, at various times, from the Canadian border to Mexico. The fact that the airplane must operate at extreme altitudes because of the high mountains—higher than those of the eastern states—adds a trying factor; in fact, with certain lower-performance aircraft, it is an impossible one. While the West Coast is generally blessed with good weather, when it is bad, it is *very* bad!

Descending through icy decks should be done as fast as possible without being wild and panicky about it. Ice raises the landing speed, covers windshields, and makes a low approach tougher. We don't want to get any more ice than necessary during the descent.

ATC will often hold or dawdle you about the icy deck, and if they do it's time to tell them that you are getting too much ice and you don't want to sit in the deck any longer than you have to. They'll generally help if you let them know you have a problem.

It's important, then, to know if cloud decks are air-mass or frontal, and what the tops are.

FRONTS AND ICE

Knowing the front locations is the most important factor in foreseeing ice, but there are other hints too.

If a cloud deck exists and the surface reports show a good solid wind of at least ten knots, not a light variable wind, from a direction we normally associate with good weather, like NW in the East Coast area, we can feel pretty certain that the clouds are air-mass and not frontal.

If we are on top it should be clear above us, or perhaps we may have only a very high, thin cirrus deck left over from the system to the east.

If, however, thicker high clouds are overhead and we are between layers, we'd better check to be certain that something isn't stirring in the general weather picture.

The first thing to do is look westward, or in the direction away from the weather movement. If these high clouds thin and decrease in that direction, with perhaps blue sky peeking through and a general clearing "feeling," then the clouds above us are probably leftovers from the past weather system.

If, however, it looks just as cloudy to the west, we must be suspicious of another weather system moving in, or the one to the east being stalled, or perhaps an occlusion bending back.

With this gloomy outlook it's time to check surface reports in all directions and try to get any new or revised forecasts. Once I carried blank maps in my flight kit and I would draw up a crude weather map in flight from weather reports received in the air. I could quickly see a change in wind circulation and begin to notice a change in weather patterns. This proved to be useful in a number of ways, like seeing a trend for deteriorating weather or for better weather; and it was excellent training to develop a deeper awareness of what weather does.

Air-mass clouds may have snow showers, but they don't have large areas of steady snow. Steady snow or rain indicates that something more extensive than air-mass weather is in progress. This we can see from surface-weather reports. An exception is in mountains with unstable air being lifted. Then the snow might be steady with very low ceilings on the upwind side, but just snow showers elsewhere. The steady snow is the result of the wind constantly lifting the air up the mountain.

When we fly fronts that have ice in them, we need the proper equipment. This time the equipment is the airplane itself. It should have the ability to climb to a fairly high level without much trouble, 15,000 feet for example, and it should have a good enough cruising range to give a wide choice of alternate airports.

NOT ALWAYS IN CLOUDS ON INSTRUMENTS

We have to define something before we can go much further in ice flying. That something is the difference between being on instruments in a cloud and being on instruments but not actually in solid clouds. Sounds a little complicated. It's like this, for an extreme: You are flying in dense haze or smoke with a lower cloud deck that obscures the earth. In this condition you are not really flying in a cloud, but you can't see anything, and so you have to fly by instruments. Now take this condition and make the haze snow, and you have the same thing: You are not really in a cloud, yet you are on instruments. This occurs very often in winter instrument flying. It's not a question of cloud density; it's just that you are either below a cloud deck or between layers of clouds in snow or haze. Whenever this occurs you do not get ice! The only time you get ice is in an actual cloud (except for freezing rain). You can fly at times for hours in such cloudless instrument conditions and not get any ice even though it's below freezing.

WARM FRONT

The structure of a warm front, and any condition that involves warm air overrunning colder air, is something like this, taking it from the ground up as if you were taking off and climbing out of a field. At first there probably will be a low cloud deck, its base at perhaps 500 to 1000 feet. This deck is caused by the precipitation falling into the colder air, and its thickness will depend on how long the precipitation has been falling and how close the front actually is. It may be scattered and only 1000 feet thick, or it may be solid and 8000 feet thick. It will have ice in it—probably light rime ice, although it can be moderate at times. As you climb through this layer it will be snowing and the windshield will get ice on it and so will the wings and other parts of the airplane, and you'll need alcohol and all the rest. The wingtips will appear fuzzy and you can see the shredded cloud slipping over the wing. Then suddenly you notice that the fuzziness on the wingtip has gone, and if you look closely you'll notice that ice has stopped forming. You are still on instruments and in snow, but you have gotten on top of the lower deck. Now you are between layers. Staying here you will not get any more ice, just snow, which doesn't bother you because you are not actually in a cloud. The air will generally be smooth. Occasionally you may feel a slight bump and if you look out you may see a slight fuzziness on the wings and realize you have picked up a little ice. What happened was that you went through a portion of cloud. In doing so you probably encountered temperatures that were fairly close to the freezing point. This little cloud is just a piece floating around in the mess trying to form, but it hasn't got quite the right formula, and the water vapor would rather go directly into snow. The lower the air temperature the truer this will be—that is, the colder the air, the less cloud. Thus we can form one rule that says, "If the temperatures are near freezing, warm fronts are dangerous things and should be considered very carefully before being flown."

At higher levels in this vertical cross section we may find another cloud deck, but it will be high, 14,000 feet or more, generally composed of ice crystals and having only light rime ice. If this deck has more than light ice you are probably in the slope of the frontal surface. Then go down to get back in the unclouded snow, or if the airplane climbs well, go up to where it's too cold for much ice. This will be above 18,000 feet, perhaps 25,000 feet.

It's wise, also, to keep the surface and lower-level temperatures in mind and know what they are. Sometimes it's easy to descend into a region of above-freezing temperatures and, of course, no ice. It's pretty silly to sit up high fighting ice when one could be lower in above-freezing temperatures.

Sometimes this other cloud deck actually doesn't exist. I remember one time flying in moderate snow at 8000 feet. I was on instruments but not actually in a cloud—and for curiosity's sake I climbed on top. The top was 25,000 feet and at that altitude there was only blue sky above. During the entire climb the airplane never went through any actual cloud, but once I was on top, looking down, the stuff below seemed to be definite cloud and it looked like the top of a stratocu deck. Upon letting down into it again I found only snow and haze—no cloud and ice!

Now as the frontal surface approaches, cloud becomes more frequent, and if there is a top deck it will eventually merge with the lower deck and there will be a fairly thick cloud deck. Here is where the front becomes interesting, because here is the ice problem—in that wall of cloud. The lower the temperature the less ice, but just for meanness' sake Nature fixed it so that you can get ice at pretty low temperatures. It's the moisture content that really counts, and that you'll have to have gotten from the meteorologist before you took off. He can look at the radiosonde observations and get a pretty good idea of how much moisture there is aloft.

Generally, the distance through this walled area isn't very great, but it can be far enough to give a load of ice that's very troublesome. The easy way is to go on top of all of it, climbing clear on up until it's CAVU above. Unfortunately, the tops of some of these conditions are very high, in the 30,000s, and you need a high performance airplane to do that. The more laborious way is to barge on and see what happens, but first the pilot should be pretty certain that he is not flying along the front, but through it via the shortest distance. This information can be obtained from the weather map before takeoff by studying the location of the fronts and the probable direction you'll go through.

FISHING TO GET OUT OF ICE

In flying up to the front you've stayed at a level that is pretty much cloud-free. That comes from old Rule Number One, about doing something about ice when it first forms. If you are flying in snow without ice and then ice suddenly forms, you start a fishing expedition by climbing if low and perhaps descending if high, carefully watching the wings during the climb or descent until you're out of cloud again. This climbing and descending will depend on the temperatures. If it's very cold it will only take a little altitude change to get out of the cloud—perhaps as little as 500 feet. As the temperature increases that cloud will get thicker and perhaps run into thousands of feet.

When we learned these things, and experimented with them, it was the era of DC2s and DC3s. There was very little ATC anywhere, so one could change altitude and fish up and down without asking for or getting a clearance—a happy time. Today, of course, we need ATC clearances before changing altitude—unless it's an emergency, and then we must announce it as soon as possible.

Now as the frontal surface is reached, climbing or descending will be a waste of time because things are all pretty much cloud, unless we know that there are above-freezing temperatures below. The pilot ought to know where the frontal surface is apt to be, so that he will not be climbing or descending when it is useless to do so, and by the same token he will not just sit at one altitude, in ice-producing cloud, before he gets to the front. All this means a careful study of the weather map and weather reports before takeoff, so that the pilot can decide where the front is, what its past movement has been, and where it will probably be when he gets out there.

Another way of telling when the front is reached is by the precipitation: It will probably increase considerably and also the clouds will have more turbulence in them. If near the location you decided upon before takeoff the precipitation becomes heavier, the air rougher, and the cloud thicker, you have reached the front.

Now it's a question of going through and hoping it isn't too far. The best out now (an out is something no sane airman takes off without) is a 180 degree turn. You can poke on in there and start getting ice. You know what's behind you and know where you can fly to keep out of ice—now it's a question of deciding how much ice you are willing to take on before you turn around. This amount must be divided in half, because if you go in and then decide to turn around, you'll have to go back through that much. Most times it isn't too far through, but it's something that makes you sit pretty much on the edge of the seat. Once through, there will be an abrupt change of some sort, the turbulence will stop, the precipitation will let up, or you may break out between very definite layers or on top. Once a pilot feels he's through the front he can start fishing again for a top, a layer, or something ice-free.

TAKING OFF IN A FRONT

It can be dangerous to take off when a front is very close to the station. You are apt to climb right into a mess that you don't know much about. It's best to wait until it's passed and then sneak up on it knowing a little of the stuff it's made of.

LEARNING TIME

All this talk has been about fronts that have below-freezing temperatures throughout. In spring and fall, however, fronts have above-freezing temperatures in various places, and while this sometimes helps, sometimes it makes matters worse.

If there are above-freezing temperatures in lower levels, it's a good time to play a little. You know then that your out is to descend and melt off the ice. It's a good day to fly high and find out what the fronts are made of. Before takeoff a close study of the weather maps and reports will give you a good picture of what is happening and where the above-freezing conditions are. Then you can go play.

But above-freezing temperatures can also be dangerous. Suppose the overrunning air is above freezing and it's overrunning below-freezing air. The precipitation will be rain, except in the lower levels. Then it will be freezing rain, and that's dynamite.

OROGRAPHIC EFFECT AGAIN

The thing that complicates all the front business is local effect—orographic lifting. With the extra help of air being lifted up mountain slopes by the wind, the air can hold lots more ice. When air is lifted that way you can get severe icing at very low temperatures. Take a cold front over the Allegheny or Cascade Mountains. It collects lots of extra water that is pushed up. That means more ice.

These orographic effects are peculiar to various regions and unless you know the region they must be guarded against. Here is where the meteorologist comes in, and a pilot, in his bag of questions, must always have one for him about the orographic effect. Always be extra cautious with ice and fronts in mountains.

COLD FRONT

In flying fronts the problems are pretty much the same whether the front is cold, warm, or occluded. A cold front is more violent, quicker to get through, rougher, and has more ice. The warm and occluded fronts are a little slower, but cover a larger area and make you sweat longer.

Ahead of a cold front there is a high deck, with some layers below, but it's generally easy to maintain a position out of the clouds and of course out of the ice.

As we fly into the front, ice will be plentiful. Cold fronts are unstable, with air being lifted quickly and therefore carrying considerable moisture. The ice can accumulate fast and heavily on the aircraft.

The distance through this area is relatively short, however. The danger, or rather the mistake often made in passing through a winter cold front, is in staying in the stratocu deck that forms behind it. The front itself may have been passed through, but the airplane remains on instruments getting ice. Actually a couple of thousand feet higher might well put the airplane on top, in the clear.

Cold fronts can be flown higher than warm fronts to avoid ice. The lower the temperature the less ice, and so up high will help. Then, when the actual solid part of the front has been traversed, there's a better chance that the airplane will be on top of the lower clouds on the backside of the front.

But flying weather of this type is like learning to play the piano—it takes time. A pilot must study the weather carefully and then begin crawling before he walks. There are certain days when he can poke into these things and see what they are made of and still leave an out or two. Take, for example, when overrunning air causes an altostratus deck—the weather is simply high overcast, with the weather at the field good. That's the day to go on up there and poke into that altostratus and fly a while toward the frontal surface, then turn around and come on out and mull it over. When any opportunity like that presents itself, it should be used.

We've flown through ice, made a letdown, and now have the runway in view. The flight is about over and we've got it made.

We really haven't, of course, because that runway may be ice-covered. Now what? First, we land short to have as much runway as possible for braking—but not so short that we undershoot.

Once on the ground we want to try the braking gingerly and see what we have. We want to do it early in the rollout. The best way is the on-and-off method we talked about before.

There's always a question about how far to go with aerodynamic braking—that is, using flaps during the landing roll, or trying to keep the nosewheel up in the air and the tail down, so that the entire airplane is creating drag and slowing one down.

It's a known, established fact, of course, that the fastest braking comes from the wheels in contact with the ground. This means that one should get the nosewheel on the ground—gently, the flaps up so that lift is lost, and the airplane's weight on the wheels where the brakes can take over. This is the best way to stop—and remember the on-off-on-off brake technique.

But if the runway is sheet ice there may not be any brakes

at all until one is moving quite slowly, and braking will be marginal then. So it would seem that aerodynamic braking would be best; that is, we should get as much as possible from it. But if a runway is that slick, then we really don't have any business landing on it. We should go elsewhere if possible.

The facts say to get the weight on the wheels and get stopped, but there may be a few times when this won't work. We cannot make a rule. This is where judgment, backed with knowledge, takes over. Pilots, I hope, will always have to make judgments; it wouldn't be a very interesting game if they didn't.

But let's say we are well slowed down on the runway and feeling pretty good about it. Now another little hazard creeps in: turning off the runway. It's often tempting to turn off as soon as possible if there is landing traffic behind us that we don't want to hold up.

But too fast a turnoff on a slick surface may find us going sideways into a snow bank. One must proceed slowly on icy surfaces. Generally speaking you are rolling along at a faster rate than you really think. When feeling you are doing 30 miles per hour, you're probably doing 50. Turn off the runway onto a taxi strip slowly.

And as we park the airplane it's important to be certain it's not on a slick surface that will allow it to move after we've left it.

Surprisingly, too, pilots sometimes hurt themselves when getting out of the airplane and setting foot on an icy surface. Our minds are apt to be so full of reflecting on the flight and landing just made that we don't get mentally back to earth soon enough to think about such details as ice on the ground. An undignified fall with a broken bone or two would be too bad, especially after flying safely through all that weather and ice!

14 Taking Off in Bad Weather

Now LET'S TALK ABOUT the three parts of a weather flight, takeoff, en route, and landing.

We have studied the weather in early chapters. We've made a flight plan, picked a surefire alternate, and decided on fuel.

The airplane has been preflighted. It's ice-free and the Pitot static ports are free of any obstructions.

The charts we need are arranged in order and placed where we can get them easily in flight. We've studied them to see what the routings may be, including departure from the terminal, en route and arrival at the destination.

There's a pad and paper handy—perhaps a knee pad. We are wearing a boom mike if we're lucky enough to have one.

We do our normal checklist and check each radio and set it for the stations required for departure. If there's a local broadcast ATIS (Automatic Terminal Information Service) that tells which runway is being used and departure information that includes altimeter setting, we've listened to it.

ALTIMETER SETTING

If there isn't this service we get the wind and altimeter setting from the tower or FSS. Altimeter setting is very important and should be set before takeoff. Don't just set the altimeter to the field elevation and think that's okay, but obtain and use the altimeter setting. This checks the altimeter's accuracy by noting if it reads the field elevation when set. If it doesn't, then our altimeter's accuracy is suspect, and we'd better dig deeper. First, get a repeat of the setting and if that doesn't correct the problem it's time to go back to the shop.

If there's an omni check on the field the ship's receivers should be tested for accuracy.

When checking the ADF, be certain it's checked in all its functions, that is, antenna, BFO, manual loop, and finally that it points to the station when in automatic.

BE PREPARED

Unless there's a directional beacon that is part of the departure procedure, and this is rarely so, tune the ADF to the outer locator of the ILS runway being used for landing. This is an emergency item in case something serious happens right after takeoff, such as a fire, and you want back in fast. You'd know which way to head as you tuned the ILS and told the tower that you were in an emergency and needed an immediate return, and what you were doing.

It's a far-fetched possibility, but why not be prepared? This also makes it important to check which runway is being used for landing, what the ILS frequency is for it, and the inbound heading. The inbound ATIS can be checked for this, or you can listen to tower arrivals.

LET'S GO

Now we start up and call the tower for taxi and airways clearance.

As we taxi out we get the pitot heat on if it's a cold day. If it's a warm day put it on just before takeoff.

During taxi, check to see that the directional gyro and turn indicator are working and indicating turns in the proper direction.

When the tower calls with the ATC clearance be ready to copy it on a piece of paper—the knee pad again. Never leave it to memory. After it's copied read it back to the tower.

After that's done let's absorb the clearance. First, mentally picture the route. Be certain the radios are set to the stations and radials required. Then firmly implant the altitudes in mind.

Most clearances request changing to some frequency for departure control after takeoff. Make a firm mental note of this frequency. Write it as part of the clearance. Also, once in the air, when new communications frequencies are given, write them down. In this way you'll always have the last frequency. If, then, you call on the new frequency and there isn't any response, you'll know what frequency to go back to. There's nothing more frustrating than to tune a frequency, discover it's a dud and then ask yourself, "What the devil was that frequency I just left?"

But let's not be hurried about changing to departure control. There's a tendency for pilots to feel they must change as soon as the wheels leave the ground. Well, the most important thing right then isn't departure control, it's flying the airplane! Before changing frequencies and talking on the radio let's get squared away, climbing, and with things set properly. A co-pilot, of course, makes all this easier and faster. But always, as in everything we do, the first thing is to fly the airplane.

Good use of radios will make frequency changes easy if we plan in advance. Since most airplanes have at least two communications radios, one can be set on the tower and the other on the departure control frequency. After takeoff, when they say to change to departure, simply flipping the control panel switch to the other receiver gets it done with a minimum of distraction.

The newest radios have fancy storage capability and one can store three or more frequencies and call the one wanted into use by flipping a switch or pushing a button.

These are useful in cruise, too; when a new ATC frequency is called for, the old one is still stored and can be recalled in case the new frequency doesn't seem to raise anyone.

As we said, mentally go over the route you'll follow after takeoff. This means to visualize the takeoff and departure. This should be a quick review of the basic stuff like, "Lift off, up gear, climb at —— indicated. Tell tower I'm changing to departure control. Up flaps at ——, set power. Right turn to one five zero. Level off at three thousand," and so on.

The point is to mentally preplan, to know in advance. Don't just fling yourself in the air and then begin to think what to do. It's all part of the important rule about flying well ahead of the airplane.

DON'T BE BASHFUL!

Of course the tower departure control or ATC may suddenly change it all and you have to be quick and flexible. Again write it down, visualize it, and if there is the slightest doubt, ask ATC for clarification. Don't be bashful! The neat cockpit with handy charts is also important here. It makes it that much easier to look up some intersection or omni that you've never heard of. There isn't anything wrong, either, with asking ATC where the heck the intersection or station is, if you can't find it.

As we tune or retune radios be certain to identify the station and have the identification sink in. This is very important. There have been accidents because a pilot was working one station thinking it was another—and this error is so simple to avoid.

It is very easy to become complacent about this because the numbers on radio dials are clear and there isn't any fishing and tuning; simply turn the dial to the number and that should be the station. But mistakes can be made: you looked at the charts for a frequency and set it up, but in the charts confusion you read the frequency for a different station; frequencies are changed and we may have missed the NOTAM, or not have the latest chart revision in hand. All these, and more, have caused wrong tuning and accidents. Even our home station, which we know so well and, perhaps, are looking at right over the nose, should be identified. Identification must be a habit!

Altitudes are also tremendously important. The ones cleared to should be written down or placed on an altitude reminder. Altitude reminders, on airline and many corporate aircraft, have warning sounds that alert the pilot he is approaching, or getting away from, the preset altitude. These are good gadgets, but the pilot still should keep his altitudes firmly in mind and not count on the gadget alone to do the job. Altitude reminders have settable numbers mounted on the control wheel hub or instrument panel. It's a substitute for writing them down—a good one, too.

Altitude is doubly important because there isn't any double check. In most cases ATC's radar is watching our track and can catch us going the wrong way, but until altitude-coded transponders are universal there isn't any double check on altitude; it's all up to the pilot.

He should anticipate altitudes. If he's cleared down to 6000 feet, at 7000 feet he ought to be talking to himself saying, "I'm leaving seven, now level off at six." In other words, climbing or descending, be at least 1000 feet ahead of the airplane. If it's a fast jet descent, be 5000 feet ahead.

When ATC clears us to a new altitude, we should get right at it and leave the present altitude as soon as possible, unless of course ATC says we can leave at our own discretion. I get a creepy feeling when ATC says to leave a certain altitude and then the pilot fiddles along for some time, or makes a little halfhearted effort at some minuscule rate that vacilates be-tween 100 fpm and 300 fpm. Man, get at it! I get that creepy feeling because I wonder if the guy they've cleared into my altitude has started toward it promptly and perhaps too fast, and will get there before we leave! When cleared one should descend or climb smoothly, but promptly. If, for some reason, such as waiting to reduce speed, one cannot, ATC should be informed.

OFF WE GO

The tower clears us for takeoff and we swing into position. If we are departing into low ceilings, we hesitate a moment and let the gyros settle on the runway heading. Check the compass too and see if it agrees with the runway heading.

If the time from start-up to takeoff is short, delay the take-off, in the warm-up area, for five to ten minutes to be sure all gyros are up to speed.

If it's night, be certain the cockpit lights are set to the best level for you, and that your eyes have become accommodated to that level.

As we start the takeoff we make a last-minute check of these few items: carburetor heat off, fuel on fullest tank, flaps where they should be, propeller in low pitch, Pitot heat on. The regular check list should have prepared us well, but it doesn't harm to have a few, most important, items that can be checked almost in one glance. My last check in a 747 takes a second or two as I look at fuel quantity, pumps on, flaps at takeoff and spoilers down—and this is after we've completed a checklist. It's a habit pattern that insures that the things that could cause trouble on takeoff are set okay.

IN THE STUFF QUICK

We should have in mind that once we lift the nose and leave the ground we'll be on instruments. It's no time to worry about flying VFR first and then transitioning to instruments; be ready at once to be all on instruments. If there's good ceiling and visibility then, of course, we want to keep an eye out for traffic, but let's suppose the ceiling is low, 200 feet or less. Also most night departures, low ceiling or not, require reference to instruments more than to an outside horizon that's poorly defined, if at all.

Going on instruments means going to attitude flying immediately; it means looking at the Artificial Horizon. With a certain amount of nose-up on the bar, at takeoff or climb power, we must be going up. So that's the first step: Fly attitude.

If the wings are level we are going straight, and if one is down we are turning—unless we're handling the rudder like a new student, which we shouldn't be—so the position of the wings on the artificial horizon is the other thing we look at.

With wings level and the nose where it produces climb, we are on a straight course and climbing. It's really very simple.

If we have to turn, we lower a wing: 30 degrees is enough. ATC doesn't expect more, and making a steeper turn means more up-elevator and more concentration that may divert you from other tasks. Keep it simple.

We first set the attitudes we want and then cross-check the airspeed, heading, vertical speed, and altimeter. Close to the ground the pressure instruments give erroneous readings. Watch the vertical speed and altimeter just as you take off some clear day. Chances are the vertical speed will show a descent and the altimeter will start down until you are 50 feet or so above the ground.

Once away from the ground the pressure instruments settle down, but even then attitude is the primary guidance and the other instruments a cross-check. This cross-check is frequent, however, as a part of our constant scanning process.

HOW ABOUT THE WEATHER?

We are off and away, but so far we've talked about flying the airplane; so let's go back now and talk about the weather part of the takeoff.

Before takeoff, if we think it's cold enough for ice, we should have the propellers slicked up with alcohol or hot with heat. The pitot heat is on, as we've said. If we are flying a jet, engine anti-ice is on. With piston engines we'll clean out the carburetors with heat just before takeoff and then remove the heat as we start to roll. But we must watch for engine icing conditions once in the air.

Before takeoff, check the wind direction and velocity. We learn any crosswind we may have to compensate for. The velocity says not only how fast we'll get in the air, but if it's gusty, it tells us if we can expect turbulence when in the air. And a thought about the wind just above the ground will remind us of any shear possibility.

The runway surface is of interest. If it's winter and the runway icy, our ability to stop if the takeoff is aborted will be materially reduced. The V_1 figure is out the window if the runway's icy. It isn't much better if the runway is wet. The slick surface will make stopping difficult.

The combination of a strongish crosswind and icy runway calls for care in taking off. Until enough airspeed is obtained to give rudder response, we're dependent on nosewheel or tailwheel steering. Neither has much traction on ice, and so the beginning of the takeoff is a little dicey, as the English say.

There's always an anxious moment, but this can be improved by getting lined up on the runway before starting the takeoff. Swinging onto an icy runway on a running takeoff is an almost sure way to be partially out of control right from the start. Getting lined up, stopped, and all squared away before starting to roll is the way to do it. Don't let the tower push you into a quick takeoff just because they are trying to move traffic faster.

Snow and slush will increase the takeoff run dramatically, and should be taken into account. Airlines don't take off if slush is over one half inch deep.

Standing water in deep puddles on the runway will prolong the takeoff run, though not as much as slush.

Any takeoff, even in very reduced visibility, is started visually. We need something to grasp with our eye. If the visibility is one quarter mile or less we need a good centerline stripe or, better, centerline lights. Without these, the visibility must be sufficient to see the runway far enough ahead for guidance.

ONCE IN THE AIR

Keep the runway heading well in mind and once in the air fly that heading until a turn is required.

Both landing and taking off in snow we find the air choppy and turbulent for approximately the first 1000 feet. This is almost always true in snow because of an unstable condition caused by the release of heat during nature's snowmaking process. It isn't dangerous turbulence, but it makes an instrument approach or a precise departure procedure more difficult.

THUNDERSTORMS AGAIN

In summer, with thunderstorms in the area, we think in other ways. Thunderstorms mean turbulence and the possibility of sudden wind shifts. They mean heavy precipitation and downdrafts. All this during takeoff, at low speeds and close to the ground, makes us quite vulnerable. If storms are very close, it's best to sit on the ground until we have some maneuvering room to get in good flying condition and avoid them.

With storms in the area, when we take off, these are the things to think about:

First, if radar-equipped, have it warmed up and ready for action. Most equipment has a "standby" position in which the equipment is warming up, but isn't actually putting out signals until the control is flipped to an "on" position where the antenna begins to search and the tube lights up.

We don't want it on in the ramp area, where fueling equipment or other airplanes are nearby. How dangerous this is is subject to debate, but the "standby" position definitely isn't dangerous and should be used well in advance of takeoff, because radar equipment takes a long time to warm up and we may want it quickly. It can be turned from "standby" to "on" before the takeoff roll starts.

A pilot needs help if he's using radar, because he can't have his eyes on the radar scope, trying to interpret the blips, and fly the airplane too, especially near the ground.

He needs someone to study the scope and tell him where cells are, and how much he needs to turn to miss them while he concentrates on flying; or else he can watch the scope and have his copilot–helper do the flying at his command.

Before takeoff with thunderstorms in the area, a good look for cells and their locations can be made with radar and an idea developed as to where to go and not go after takeoff.

The flight path one would like to fly, then, after takeoff can be worked out with ATC before one is airborne. It's so much calmer to do things like that on the ground than in the air, on instruments and busy.

While we are trying to duck thunderstorm cells, ATC has to be kept advised and advance permission for turns obtained if possible. This is especially true in a terminal area.

Aside from radar we have to think about the wind during takeoff and how close the storms might be: It could be possible to get a hefty, gusty wind shift during takeoff.

Suppose we are taking off southward into a strongish wind. A thunderstorm arrives and snaps the wind to northwest, our takeoff run is lengthened, and once in the air, if we get there, we'll probably be in downdrafts that will make the climb sluggish at best.

It's a silly time to be taking off, of course, but sometimes it is difficult to tell just where the storms are if lower clouds hide the actual storm clouds.

Once in the air we may get some pretty wild turbulence. It's a difficult situation because our speed is low and we start to climb, and a severe bump might get us close to stall. A wing can be pushed down at the same time.

What we want is to get up to a good workable airspeed as

soon as possible; but, of course, we don't want to shove the nose down and fly back into the ground either. It all takes care and judgment.

We should have in mind a direction to run away from the storms if things get too tough. This knowledge comes from a careful study of the weather before takeoff in order to know how the weather is moving. Study the sequence reports for stations in the area—a 100-mile circle at least—to see what is actually happening.

This sort of weather is tough and it shouldn't be taken lightly.

THINKING

Now we are in the air on instruments, encased in the tiny world of our airplane, everything outside ending at the windshield. Let's not worry about relating ourselves to the outside, trying to see out there or holding on to that world we left. It's best to forget it, settle down, and watch the instruments.

We concentrate on the rather simple job of flying instruments. It is more simple the more we are relaxed. Lack of worry makes us relaxed. Uncertainty makes us worry, so let's not be uncertain; let's know what we are going to do next, be prepared, and plan ahead.

Our thinking is in layers. The top layer is flying instruments. But this is easy and doesn't require much of our thought. If we've practiced enough it's almost automatic.

The second layer is the airplane. Are we doing the right things to keep it running? Carburetor or engine heat? Power settings, pitot heat on? Fuel on proper tank, the gear and flaps cleaned up? In other words, all the things that will ensure the airplane's running. If it stops running, so does the problem, right there! So it has to run.

The third layer is navigation. What's the radial we are trying to hold on? Is the station identified? What's the altitude? Are we level or approaching it? That intersection coming up, are we tuned, identified, and set up for it? We review all the items that are a part of our navigation, including holding a heading the man on the ground may have given us to hold.

Layer four is the weather. Turbulent? Fly turbulence methods. Ice? Is all the anti- or deicing equipment being used, and what does the clearance ahead do for our getting out of the ice? Thunderstorms? Are we ready for turbulence, is the radar working and do we know a way out if we want to run?

Layer five is a quick thought toward emergencies. If I want back in how will I do it? If I cannot get back in where's the nearest place I can get in?

These layers are not precise, nor are all the items listed, but the thoughts for building your own are there. The layers intermix at times, too; but once we get the airplane and engines set we don't have to think about each thing constantly. If the weather equipment is on and working we don't have to think about it constantly. The emergency action, once it's implanted in our mind, doesn't require attention; it's there if called on.

Each layer has a priority based on how often we need to refer back to it.

As we navigate or follow radar vectors, it's wise to visualize where we are and what's under us. We shouldn't concentrate on this to the extent of neglecting anything else, but we should keep in mind roughly where we are and what we are over. This is particularly important in connection with terrain height. How high is the terrain and are we clearing it with enough margin?

The charts give minimum altitudes and we should pay attention to them. This is a part of preplanning; before takeoff we should have a good idea of where things stick up, be they TV towers or mountains.

Radar vectors for departing, en route, or landing, and the altitudes the man on the ground gives you, are supposed to keep you from hitting anything. But don't count on them 100 percent. The people on the ground aren't 100 percent perfect, and neither are you and I. There have been accidents because an airplane was vectored into high terrain. Very few, but they have occurred.

In the final summing up, the man flying the airplane is responsible for it, and missing the ground is a paramount part of the job. People on the ground seem to take over more and more of our flying, but it's still basically the pilot's responsibility. Never take anything for granted, and especially not the idea that a certain heading and altitude given from the ground will clear all obstructions!

We must, by any means possible—Omni, RNAV, ADF, DME, INS, whatever—keep up-to-date on where we are. We must not simply sit there, wandering thither and yon following the ground-given vectors, and not know where we are in relation to terrain. The controller may be giving headings, but the pilot is responsible for the navigation. Navigation is made up of many things and one of the biggest is keeping position information and knowing where we are! The heading, whether provided by compass or ground-based ATC controller, is simply directional information. Where the aircraft is, and if it is clearing all terrain, is *always* the responsibility of the pilot in command!

Weather Flying
En Route

Now WE LEVEL OFF at cruising altitude; sit back and relax a little. The intense concentration of an instrument departure is over.

This is time to look around the airplane and check things. Glance at the fuel setup, that pitot heat again, the engine heat, the deicing equipment. It's a good idea to start on one side of the flight deck and cover everything across the instrument panel to the other side.

The charts and papers from takeoff are probably mixed up, and this is a good time to get them organized and neat. Now there is an opportunity for writing down the takeoff time, the time a fuel tank was switched on, and for the other bookkeeping jobs. Knowing when fuel tanks were turned on and the time takeoff was made, along with the time over checkpoints, is important.

During all this, of course, we've been watching the weather. En route it's a clear-cut problem of being or not being in an area of instrument flying, with possible ice or thunderstorms. Turbulence can be mixed with either. It's natural, of course, in thunderstorm areas, but if it's in an icing area the ice will be fast-forming and heavy, and our problem is to get an ATC clearance out of there. Such a situation means unstable clouds, generally, and often a good way out is up. The clouds, probably embedded cumulus in a general overcast, will occasionally be visible as we bounce in and out of them between layers.

THINK AHEAD

We've talked before of the different weather conditions and how to cope with them, but what we want to talk about here are the things we think about while flying from one place to another.

Of course we are navigating, listening to the radio, working ATC, and keeping track of groundspeed and fuel usage to know if we'll get where we want to go with fuel enough remaining to go to the alternate.

As we do this our background thinking is doing the important job of keeping ahead of the airplane. This is a tremendously important part of flying—in or out of weather, but especially in weather.

What do we do to keep ahead? We have the charts for the next leg of the trip in hand. As we fly toward one omni station we have the course out from it well in mind before ever getting to the station. We know what the next station is, or intersection. We are mentally prepared for a number of courses ahead. We have the terrain and altitude clearance well in mind, too.

The time a fuel tank is due to run out is noted, so that we'll begin to think about it in advance and change it before it runs out.

But most important is the weather. We listen to broadcasts and copy sequences for our route, destination and alternate. It takes more than one hour's weather-watching to catch the trend, to see if it's doing what the forecasts said it would do. If it isn't, then what does it seem to be doing? Is it getting worse or better? So we copy weather each hour and compare it with the last hour.

Better weather, of course, isn't any problem. Worse means we have to plan for alternate action. What might that be?

First, we should check deteriorating weather against the weather at our alternate. Is it holding, can we feel confident that it is still the safe place to go if we cannot get in at our destination? If it is, we keep going to our destination, make an approach, and if we can't get in we proceed to the alternate. Just like the book says.

If the alternate is going down, too, then we need a new alternate. If our study of the big picture was complete before takeoff, any change from the forecast should be understandable. We should be well equipped to take alternate action. A good weather pilot has this alternate action in mind before takeoff. He has in mind alternate action that fits the weather setup that day. Again and again, knowing the big picture is important!

WHAT'S IT LIKE?

What's it like en route, as far as our flying condition is concerned? Well, we'll either be on instruments or not. We can be under clouds, on top of clouds, or between layers. We can be in and out of clouds.

Un-supercharged piston airplanes will be in cloud more than any other type, but even so important "instrument" flight will be between layers or on top. Clouds rarely stack up from the ground to some great altitude. When they do, it is in a front or near the low center. These are not far across, nor do they stay in one condition for long periods.

A big portion of weather flying occurs in postfrontal conditions, and that means flying on top. Prefrontal conditions will be largely contact under an overcast that may gradually squeeze down on one.

Approaching a warm front in winter we may fly on instruments for long periods in snow. There isn't ice, as we've said, unless one gets in stratus clouds formed down low (and a short climb will top those), or in the area where the higher clouds lower and meet lower clouds—and this will be near the front.

What we are saying is that most instrument flight isn't! When it is, it's because we are flying in a front or low, as we said, or we've made a bad choice of altitude—like being in a stratocu deck when we could be on top of it.

This allows us to sum up and list the conditions. We've said we'll be either on top, under, or between. In these conditions we are staying out of ice, and we should be able to see thunderstorms and avoid them. When the clouds get together we can get ice or fly into a thunderstorm. We are near the frontal surface. If we aren't, then we've just slipped into clouds that are higher or lower than the ones we've been flying through. If we understand the weather situation, then we should know whether to go up or down.

If the clouds have enveloped us and we're near the front then perhaps it's time for ice. This isn't a reason to feel all is lost. Many times there are above-freezing temperatures at a lower level. Then it's simply a matter of getting a clearance to go lower, remembering terrain clearance.

If it's cold clear to the ground, then the job is to be certain that we are above the lower stratus clouds. This means a higher altitude. But in getting this higher altitude we are not trying to get on top; we are just fishing for an area where there's little or no cloud, just snow. Trying to get out of ice by topping a warm front is a bad way to do business, unless you have a jet or very high altitude capability—20,000 feet and above. If we climb, say, from 5000 feet to 9000, we may climb into warmer, overrunning air, which has more moisture and a chance for heavy ice.

We can pick the altitude above the lower stratus clouds and go into the front knowing we'll get ice and watching to see that we haven't gotten so much that we cannot turn around and get out.

If we are crossing the frontal surface at right angles, it shouldn't be too far through; so let's be certain we are crossing it the fastest way. Let's not fly along the front; then we'd really ice up.

Thunderstorms we've talked about. Either they are air mass and we wander around them, or they are frontal and we are on instruments around them and we need radar. Without radar, don't be in there! If the radar breaks down once we're in it, be prepared for a rough ride.

Which points up that we shouldn't fly fronts—winter or summer—until we have a lot of instrument experience.

So en route we watch weather at our destination and alternate, but we also watch to see if any fronts are getting on the route.

Generally precipitation and easterly winds make us suspicious of a warm front. The frontal surface is out there somewhere where the winds go to southwest from easterly. Ahead of that is the large area of precipitation. That's where the weather is, and where we'll find imbedded thunderstorms in summer, and sometimes in winter, too. We'll find snow, and in a more or less narrow band, we'll find freezing rain. Until we're experienced we simply run away from this area, either by making an end run if we have a long flight and this is possible, or by landing and waiting it out.

We pick up a cold front more easily, since its wind shift is more dramatic, SW to NW, with heavy rain, thunderstorms, or heavy snow and snow showers. It's rough in summer or winter, with ice in winter, thunderstorms in summer. Are you equipped to take it on? Equipped in skill and knowledge and airplane and gadgets? If not, beat a retreat and wait for it to go by.

Flying en route is the crafty time. Watch the weather, destination, and alternate, and developments along the way that may mean something different on the move—big picture again. Watch the groundspeed and its relationship to fuel and distance. These, together and interrelated, are the budget, and we cannot afford to overspend.

FORCED LANDING

Suppose we are faced with a forced landing on instruments. What then? First consider the terrain under you. If there are mountains, turn so you descend parallel to them and not across. That way you have a chance of coming down in a valley and at least not banging head on into the side of a hill.

On top of clouds one often can see a wave-like pattern in the clouds below. If so, descend in the low part, the trough of the wave, and you will probably be over a valley.

Slow down to minimum airspeed, but not so slow as to stall until ready to touch. Ninety-nine chances out of 100 you'll see something before you hit and have a chance to maneuver and pick a softer place. So you want a little maneuvering speed.

Full flaps and gear down. The gear will help to decelerate as it tears off and the flaps will give the lowest speed, of course. With shoulder harness and pillows in front of you and your passengers, things may not turn out badly at all. The trick is to hit under control. It all sounds pretty desperate, but it isn't quite that bad. We hope it never happens to you, and it probably never will.

And, of course, en route flying is handling navigation and ATC. It is also, during the not-so-busy cruise portion, a chance to plan the arrival, to get out the arrival-airport area charts and study them for routes and holding points. It's the time to study the instrument-approach plate and firmly fix in mind the lowest altitude we'll go to and what the missed-approach procedure is if we don't get it.

The approach plates should be held by a handy clip on the control wheel or instrument panel so that we can see them easily and without juggling them on our lap.

With all this done we are ready for descent and landing.

16 Landing in Bad Weather

As WE APPROACH the destination, weather takes on a more realistic feeling as we relate ceiling, visibility, wind, turbulence, and runway condition to our landing.

First, we try to visualize the weather in relation to our descent. Is there ice or are there thunderstorms?

If ice, as we've said, the job is to get well prepared; heat, props clean, etc., and descend quickly to a landing before too much ice covers the airplane.

Thunderstorms we have to detour around.

Now let's look back at the basics of an approach. There are two kinds, one a precision approach with glide-slope guidance to the center of a specific runway, which is an ILS or PAR. The others don't have glide-slope guidance or precise runway lineup; they are omni or ADF and called non-precision. These approaches have higher minimums because the guidance isn't as precise. But because the minimums aren't as low, there's no reason for not taking them as seriously as an ILS.

On the contrary, since ADFs and omnis aren't as precise, we should do the best possible job so as to reduce their imprecisions to a minimum.

FLYING THE APPROACH

Courses followed exactly and altitude controlled with precision are the rule. The altitudes on the approach plate are to be used accurately, and times from final fix to the field followed closely. Times for letdown should be modified according to the groundspeed we think we are making.

There are three phases of an instrument approach:

1. The all-instrument part.
2. The transition period from instruments to visual.
3. The visual part.

The instrument part is the easiest part. This is a mechanical thing we learned when we got an instrument rating. But there are a few tricks and points worth talking over.

First of all it's important to get the airplane settled down and ready for the approach—that is, to get all the prelanding checklist items we can out of the way well in advance. We shouldn't be worrying about changing fuel tanks, getting gears down, etc., while we are trying to get on an ILS and stay there, or home on an omni or track an ADF. Have everything possible set for landing before the approach is started.

THE INSTRUMENT PART

The most important part is to keep on top of things every moment. We are talking again of scanning and keeping headings where we want them along with altitudes, descent rates, and airspeeds.

Let's remember the simple method of keeping a heading by watching our bank on the Artificial Horizon. If we never let bank go unnoticed, we will not get far off the heading. A part of this is promptly changing the heading when it should be changed. In other words if you think a heading change is needed, make it now! The idea is to keep a short rein on the entire process, and never let the airplane get far from the wanted path. Scan often—act quickly, but of course smoothly.

Glide slope on ILS, or altitude change from an omni, is mostly a matter of descent rate. A localizer needle is not flown; the directional gyro is flown to a heading and we refer to the localizer needle to see if that heading keeps us on course. In descent the vertical-speed indicator, hopefully an instantaneous type, is the key. You set up a certain rate of descent and see how that keeps you on the glide slope. If we go below, the descent was too fast, and so we get back on and try a slower rate. We bracket the glide slope using the vertical-speed instrument just as we bracket the localizer with the directional gyro.

So the two important instruments are the gyro and the vertical speed—one for course control, and the other for descent.

The horizon, showing attitude in roll or pitch, is the instrument by which we make our gyro turn or our vertical speed change. If a bank occurs the heading will change and we stop it, or bank to make a heading change back. If the horizon shows our nose down more than it has been, we bring the nose back where we want it; if we want to go down we make a forward movement which shows on the pitch part of the horizon.

We refer to other instruments too: airspeed to keep it within bounds and the altimeter to see we're where we ought to be.

We often hear arguments concerning whether we control airspeed with the elevators and rate of climb with the throttles, or vice versa. Well, it's a silly argument because what we're managing is energy and we do it the best way for the condition we're in. If we're below the glide path, our natural reaction is to pull back and get up to it. We zoom a little and get back, but if we're quite low and our zoom takes a long time, we will lose energy and slow down. Any experienced pilot knows this, and as he pulls back he automatically puts on some power. If it's a little pull up, and he has some extra speed, he will not add much power—maybe none. If he's slow and quite low he may start adding a lot of power right away because his experience, subconscious or whatever, tells him he's going to need the energy.

But stopping to think which to do, add power or pull back, just isn't the way to fly an airplane. It's a smooth coordination of whatever it takes to get the job done under present conditions and if one doesn't understand that, he'd best get out and do a lot of practicing and learning before trying to fly down an ILS looking for the bottom of the clouds at 200 feet or less!

To do it all, we scan and scan often.

In a descent from an omni or ADF we do not have glide slope. We simply leave one altitude and go to another. The object is to get down to the desired altitude as quickly as possible. This doesn't mean we need to dive for it; it simply means to get on with it. The vertical speed shows us how fast we're doing this. Obviously we don't want a descent rate that will make recovery at the desired altitude difficult. That's dangerous. What our descent rate should be depends on how much time it will take us from the final fix to the field. That says how long we have available to get down and that tells us the descent rate: If it's a minute, omni to field, and we have 500 feet to lose, a 500-foot-per-minute rate of descent will get us there on the button. But since we don't really know what speed we'll make station to field, we should add a little. If I had a 500-foot-per-minute descent rate required, I'd probably use about 700 FPM to be sure I'm down there before I get to the field.

The descent rate is also a guide for leveling off at the desired altitude. If the rate of descent is high, then we begin to level off farther above the altitude than we would with a low descent rate. We don't want to go below our target altitude. A little practice in hitting altitudes from various descent rates will soon let us know what our individual abilities are. It's good stuff to know before we start making approaches.

We want to get the approach tied down as far out as possible to reduce the need for excessive maneuvering close in. An approach is a little like a funnel. Well away from the runway the localizer is wide, and as we get close in it becomes more narrow. This creates a tendency for us to be a little less precise out near the outer marker than we are in close. The way to make a good approach is to get on course and stay there as soon as possible. A good portion of missed approaches don't happen in the last part of the approach, although it looks that way; the miss probably started way back at the beginning of the approach when the pilot didn't get on the track right away.

The wind will change as we descend and this will affect our drift and descent rate. We will have to change headings and descent rates to keep up with it, but the earlier we have these things under control, the easier it will be to pick up a change. And remember, the descent rate to stay on the glide slope is a clue to shear.

What we are trying to do all down the approach is to keep the changes as small as possible. Close in, the ILS is narrow, the omni station small, and small corrections do a lot. It's the time and place for maximum concentration.

CLOSE IN, THINGS GET TIGHT

There's a lot of pressure on the pilot as he gets in close on the approach; everything is tighter and tighter and he must watch each instrument intensely and scan the important ones frequently. This is the place where hands get sweaty. It's the place, too, where the pilot has to talk himself into being relaxed, but alert. He must feel like a good athlete, intense in concentration, but smooth in coordination and relaxed enough so that he moves freely. This comes by practicing, by getting the approach under control early, and by talking oneself into the right mental attitude.

STICK WITH IT

As we anticipate breaking out and seeing the ground we should remember one of the most important parts of an instrument approach; it is this: Stick with it!

All approaches are not to the minimum altitude before seeing the ground. Most of them are not. The ground comes into view in various ways; through breaks in clouds, or a clean break out of cloud but into poor visibility, perhaps straight down in snow. We often see the ground before we have sufficient visibility to see the airport or runway.

But seeing the ground, good old Mother Earth, whom we may not have seen for hours, doesn't mean the approach is finished. It isn't until we land and turn off the runway onto a taxi strip.

An inexperienced pilot, seeing the ground, may try to continue the approach visually, to navigate by ground reference. He thinks he knows where he is, but with poor visibility and the field not yet in sight, things look different, even around an area one thinks he knows well. Suddenly he runs into low scud or a snow shower, and the ground disappears. It's a shock!

During this ground-contact navigation he has wandered from the radio aid, but now, with visual reference suddenly gone, there's a wild attempt to get back on course with radio aid, but it's too late and the approach is missed.

When a pilot leaves the radio aid and begins to navigate visually, he probably starts to descend. It's a natural action, to get down and have a better look, and it's done unconsciously. It's a false concept and a very dangerous one too.

The altitudes for an instrument approach are set up to clear all obstructions along the approach path. It is best, by far, to follow these altitudes precisely and not be "suckered" into letting down below them because one can see the ground and get the false impression that this makes everything okay.

If there is enough ceiling the approach altitude will put you in the best position to land. If there isn't enough you shouldn't be landing anyway. Why deviate from the procedure? There isn't any good reason, so don't. I cannot stress this enough.

All this is especially important at night, even in fairly good visibility. The lack of good, solid reference can cause sensory illusion that will lead a pilot into the ground when he thinks he is doing fine. A few lights or, in daytime, a dimly seen building or tree clump, combined with a wing being down and an off-level nose position, can easily give the illusion that one is higher than he actually is.

When the runway or approach zone slopes up you think you are higher than you really are. When runway lighting is set to an intensity lower than normal you think you are higher than you are. Snow cover does the same thing and under conditions of haze, smoke, and darkness you think you are higher than you are.

These sorts of illusions have caused accidents. The answer is to trust instruments, not what you think you see. Stay with the instrument-approach procedure until the runway is in view and it's definite that you "have it made."

This is the dangerous place, when one casts aside instrument reference and uses only one's eyes for guidance. Again, and again, we must not descend below the final altitude until the runway is well in view; we can adjust our descent by reference to it, and we're certain of our landing. Even at the point when we cross the runway's end, we should be checking instruments—airspeed and especially altitude. Even with the runway in view it is possible to get too low, because our eyes play tricks, and a look at the altimeter will sometimes be startling! But if one has taken frequent quick looks inside at the altimeter during descent, even though "contact," these surprises will not happen.

During an ILS approach keep in mind how much crab angle has been necessary to stay on course. Let's say the ILS course is 50 degrees and it has taken about 60 degrees to keep on the localizer. This means when we break out and see the runway we'll have a 10-degree crab angle. The runway will come into view out the left side of the windshield. The instinctive reaction is to kick this crab out and line up with the runway. We shouldn't, because the airplane will then drift off the centerline, and it's surprising how fast this happens. If the ceiling is low we'll probably not have time enough to maneuver around and get back on the runway before we reach it. It's a missed approach.

So on the ILS, note the drift and visualize where and at what angle the runway will come into view, and then don't be too quick to kick out the drift; see how you are doing first; you may have to hold it all the way to the ground.

WHEN WE SEE AGAIN

Now we've made the approach and we break out where we should. What happens? Unfortunately there's a great tendency to push over and "go" for the runway. They call this the "duck under" maneuver. What it does is put one too low and it's dangerous.

To go back a bit: We were on instruments following the glide slope and we established a good rate of descent. But the moment we look up and see the approach lights we lose glide-slope information. The approach lights don't give any. The radio glide slope is inside on the instrument panel, not outside where we are looking. Unless there's a Visual Approach Slope Indicator, VASI, we haven't any glide-slope reference.

Looking back inside at the glide-slope indicator doesn't help because the glide slope deteriorates as we get lower.

ILS systems approved to CAT 11 minimums, generally 100 feet, have accurate glide-slope indication almost to the ground, but others are not that precise and their glide-slope accuracy deteriorates as they approach the ground. How high this happens depends on various things, but terrain has a lot to do with it. Some glide slopes are good to 50 feet, but I've seen others that dip or become erratic as high as 150 feet. So at low altitudes if the glide slope isn't a CAT 11, and the needle suddenly makes a fast and pronounced excursion off course, don't chase it; simply hang on to the rate of descent that's been taking you down the glide slope successfully and use that for a glide slope.

This is a dangerous area. Enough airplanes have landed short under these conditions to prove the point. The airplane condition at that point aggravates the situation. It is going slowly, the pilot pushes over; the airplane goes through a shear zone, it sinks and sinks to a dangerous level before the pilot can catch the cue visually, if ever. What we need is VASI on all instrument runways, but generally we don't have it. Someday, too, we may have a heads-up display, where the necessary guidance is on the windshield. This, plus improved radio glide slopes, will mean the pilot can look out at the ground and have his radio information too.

But what do we do about it now? First, remember the rate of descent you've been carrying in the approach. When you break out take occasional quick looks at the instrument to see that you are holding near this rate. If you are, the descent path should be nearly correct unless there is bad shear. Check altitude too, especially in relation to the middle marker, the amber one. On the approach plate there is a published altitude for crossing this marker. It will be around 200 feet AGL. Be certain not to be below this middle-marker altitude until after passing it.

What we've said in all this is not to trust your eyes for glide-slope guidance when visibility is low, especially at night. This applies to the VFR pilot under fairly good conditions too. We've also said that an approach doesn't end just because we can see the ground. And when we see the ground, we must kill the desire to push the nose down and duck under!

TO TOUCH THE GROUND

So we've made a proper approach and now see the runway —what's next? Actually, we continue with the descent and approach problem. We cannot get too low until the actual pavement has passed under us. While we want to get on the ground, we also don't want to try to cut it so short that we leave the wheels on the last row of approach lights. It's one of aviation's many compromises in that you don't want to land too long with the danger of going through the fence at the other end, and you don't want to land too short either.

This points up an important point about landing the airplane once the runway is securely under it. The point is to get it on the ground! This isn't the time to float along with a little excess speed trying to gently slide the airplane on the ground and impress those riding along. Even if the landing is a little rough get it on the ground where the wheels and brakes can begin to get it stopped. And in doing this we want as much runway ahead as possible in case we start aquaplaning on a water-covered surface or sliding on ice.

So there's a nice, delicate balance required in the approach. It starts with airspeed before the runway. We need enough to take care of gusts and shear, but not so much that we come screaming in over the runway and float halfway down it before we can touch down. We also don't want to land so short as to risk hitting the approach lights, but we don't want to land so long that valuable runway is flown over instead of being used for landing on and getting stopped. These are the kinds of judgments pilots make alone and without help.

Now a few points about low visibility. It's interesting to note the following table of the percent of missed approaches versus visibility as taken from records at London airport for three years:

Runway Visual Range (Feet)	Missed Approaches (Percent)
1970–2300	22.2
1620–1970	30.6
1475	40.5
1312	45.5

It simply says that it's tough to make a successful landing when you cannot see. It says there isn't any point to being a hero trying to get in under conditions that are just too tough.

These low visibilities are generally there because of fog. During this condition the visibility measuring device may only see a portion of the runway area, although some runways have RVR measurement in two and possibly three areas along the runway and give rollout area visibility values. Beyond that, where one may roll out, there could be a zero-zero fog patch. Very low visibilities are touchy and require objective analysis about whether or not it's worth trying before doing so.

GROUND FOG

There's another fog condition that deserves attention, and it doesn't have to do with an instrument approach. We can fly in clear weather, day or night, and note that our destination reports zero in ground fog. We get over the field and are surprised to see the runway and airport below us well in view.

This is simply because we are looking down through a very shallow layer of fog, rather than through it horizontally. But if we say, "Heck, I can land in that," and make an approach, we'll get an awful shock as we descend and suddenly, about the time we start to flare, go on solid instruments.

What then? Pour on the coal and get out. If it's too late for that, hold a steady descent rate and attitude until you touch and then get stopped as fast as possible and hope you don't hit something before you do.

Back to our low approach to an airport. We've gotten on the ground and stopped; is the approach over? No. The next order of business is to quickly clear the runway. There's a guy at the middle marker about to land and the tower cannot clear him until you're off the runway. So turn off as soon as possible.

In doing this remember the so-called high-speed turnoffs aren't as high-speed as they sound. If the runway is wet and we start to turn off at fairly high speed, we can be shocked to find that the nosewheel may have turned toward the taxiway, but the airplane hasn't, and the nosewheel, without much weight and traction, is just skipping sideways while the airplane is going somewhere between the runway and taxiway—which is in the boondocks! Find at what speed your airplane can make that turn on a slick surface and be down to it before trying to turn off.

THE TOUGHEST CASE

Suppose the worst happened and for some difficult-to-explain reason one got caught with no alternates, no fuel, and a fog-covered, zero-zero airport the only place to go.

I about did this in Alaska on the Aleutian chain during World War II. It wasn't zero-zero, but might as well have been. The ceiling was essentially zero and the visibility about 200 feet in snow and fog. Fortunately, I was over a good airport, Shemya, which had a long and wide runway.

I came down the ILS and carefully checked the descent rate needed to follow the glide slope. I flew the ILS as tight as possible. When we were under 200 feet and the glide slope started to wiggle and deteriorate I just held the rate of descent until we touched; then, wheel forward—a B17—slammed on the brakes and tried to stay on the localizer and get stopped before hitting something. We made it.

What really impresses one is that with near-zero visibility you really don't see! A few fuzzy lights aren't enough for guidance and you are helpless and feel helpless. This feeling can occur with visibilities up to 1000 feet or so, depending on how fast you land. It's an excellent lesson to make one realize that there are human limits and we shouldn't try to exceed them; too many things happen too fast.

I've made many approaches in a 747 to 100 feet with 400

meters visibility. It's done on automatic pilot and is easy because you have time to check all phases of the approach constantly, while a bunch of transistors and servos do the manual labor of keeping on course. Without this capability a pilot flying very low approaches by hand either will be concentrating so hard on flying that he will not have time to check other important things like altitude and instruments that tell the flight's entire condition, or, if he's checking those things, he will not have time to keep on course and complete the approach. He cannot do it unless he's a superman, and while I've known the best of pilots, none of them are supermen. One reason they are the best is that they know when to admit something cannot be done safely!

If one gets into a zero-zero fix, one flies the ILS as best one can. Make one or two runs down to 200 feet checking the drift and descent rate. On the for-keeps approach, hold the descent rate after passing the middle marker right to the deck. It will be low enough, 500 to 700 feet per minute, to make a fair landing. Then try to steer down the localizer and get stopped as soon as possible and hope for the best. You might as well stick with the instruments, because looking out will not help at all, while instruments will.

Better by far, however, is to stay out of such situations!

17

Teaching Yourself to Fly Weather

WITH THE IDEA that one crawls before walking, we can teach ourselves to fly weather. It's a progressive process and we can set up our own program.

In the work of getting an instrument rating we made practice IFR cross-country flights, getting clearances, making radio reports, navigating, gathering weather, and all the rest, even though we weren't flying actual bad weather. This is the way to start our "home study" weather program.

Each day, in our advancing times, the complexities of air-traffic control, routes, and communications grow, so that all the experience we can get in this area is important.

If, on each flight, VFR or IFR, we are on a flight plan, doing all the work required, we will become facile with this part of the job and do it smoothly, almost automatically. Once this has become an easy task, we will have time to think about the weather.

WHERE'S THE EMPHASIS?

The balance in flying between ATC and its communications versus the actual weather has tipped toward ATC as the major problem. I am certain that weather accidents have occurred because pilots became so engrossed with ATC and some altitude or route ATC gave them that they got in trouble with ice or thunderstorms because their attention wasn't on the ice or thunderstorm as the primary problem. We tend to be more afraid of authority than of weather, and its legal experts.

This, of course, is silly. If the route or altitude assigned doesn't fit, it's time to tell ATC that you want something different. If we are battling a difficult situation and ATC keeps pressing us with complex routings or unreasonable requests, we should be prepared to tell them that we have urgent weather problems and need help, not hindrance. They will help if at all possible.

All this comes easier if we are prepared by being able to handle the ATC complexities, by having a well-organized cockpit, by knowing the regulations, and by having a boom mike to make transmission a simple matter that doesn't require taking a hand from the flying job.

It is difficult to find just the weather we want for practice, but ATC is always there for practice. We should do as much as possible.

LEARNING THE WEATHER

Now let's get on with learning to fly weather.

The actual weather part we will sneak up on by flying a little at first, more as we gain experience.

Following is a step-by-step method. These steps are guides and one's own judgment will vary them as one appraises his growing ability and degree of comfort in different stages of weather.

The idea is to fly weather with safeguards that relate to our experience. After we've flown the first step's conditions enough to feel comfortable, we can take on a little more as in step two, and so on. The steps are:

1. Fly good weather to good weather on top.
2. Bad to good.
3. Good to bad.
4. Bad en route.
5. Thunderstorms.

Now, let's talk about them.

STEP ONE

The first step, *good to good,* means we leave a point that has broken clouds or better and fly toward a point that has the same condition and is forecast to remain that way or to improve. The ceilings should be 1000 feet or higher, the tops 7000 feet or lower. No fronts moving toward destination. Temperatures above freezing and no thunderstorms.

This flight will mean climbing up through clouds to on top, and descending through them at destination. It will require weather-watching to be certain things stay as we want them and, of course, a flight plan and ATC.

This condition will generally be found in the stratocu cloud decks behind a low passage. It gives an opportunity to take on more or less weather depending how close behind the low, or front, we take off.

At first we might wait a day after the low passage for the stratocu deck to become thinner. Later, as we gain experience, we can take off closer and closer to the departing weather until we are taking off with a ceiling that's quite low, climbing to a top that's quite high, or flying on instruments to a destination that's overcast.

STEP TWO

This leads into step two: *bad to good*. It's simply a continuation of the first step, but departing closer to the low and front until we are taking off just after its passage. We should be careful not to take off before the front has passed or while it's passing, because this can be a wild experience.

The key thing in both these steps is that we are always flying toward good weather. We want to be certain it's forecast to stay that way, too: No fronts moving in, as there might be if one took off from an airport that just had a warm-front passage with weather improvement, and then flew across the warm sector of the low. In this case one should realize that out west somewhere there's bound to be a cold front coming along that we don't want to get involved with. When starting these first steps, it's best to take off after a cold front has passed; then there shouldn't be any more fronts for quite a long distance.

There are special situations, like the Los Angeles basin area, that are excellent for bad-to-good flight experience. The frequent low stratus allows for an instrument departure and a climb to on top, where it's CAVU, and then a flight to someplace on the desert like Palmdale where the weather is good. And generally this can be done without running into the problem of fronts.

We can back up and start these two steps over, but with some ice. The first sight of a smear of ice across the windshield is quite interesting if there's a way out nearby. We can do this by climbing through a thinnish, below-freezing stratocu deck, 4000 or 5000 feet thick, that has a good base, 1000 feet or better, at both departure and destination.

As we get more accustomed to ice we can take this deck a little thicker and a little lower.

A good method to learn about ice is to fly in a stratocu-deck that has ice, and at the same time a freezing level above the ground, so one has the chance to go down and melt any ice off. The freezing level must be above the ground enough for 2000 feet clearance or so. This condition is often possible in the spring and fall.

STEP THREE

Good to bad. This means to fly toward a destination that has weather approaching. The departure should have solidly good weather. Doing this, we can go to an area of low ceilings, shoot a low approach, and if there's trouble, turn around and go back to where it's good.

Plan, at first, to get to the destination well ahead of the weather, but be prepared for surprises. It may move faster than expected. You'll learn, this way, how foolish it is to fly to a destination that calls for deteriorating weather, without a good out to run for.

As experience is gained in this step, one can fly to destinations with worse and worse weather. The key is to have that out—in this case good weather behind.

This step should be tackled first without any ice, and then we gradually work into ice as we've done in the previous steps.

STEP FOUR

Bad en route. This means we'll take on a situation that has a good destination, but something in between that is tougher than just a stratocu deck. It will be some kind of front.

To start this we should be without ice and thunderstorms, just as we begin each step. The destination should be forecast good, and to stay that way or improve. Any front should be well past the destination before we take off. The takeoff point should be 1000 feet or better and forecast to stay that way for at least two hours after takeoff.

We progress in this step by taking off with the weather closer to our departure point. This requires thinking about a takeoff alternate—where to go if you must return and the takeoff airport has gone below limits.

We develop this step until we are taking off with bad weather, flying through it en route, and landing at a point that has recently cleared. At this point of progress you have reached a pretty sophisticated level in weather piloting. But there still remain thunderstorms.

STEP FIVE

To begin *thunderstorms* we should fly with only air-mass types en route. They should be forecast as scattered so that we can wander around them and look them over from a safe vantage point. There shouldn't be any fronts forecast within 500 miles of our route or destination. We should have lots of fuel. The destination may be covered by a heavy shower when we get there, and we want enough fuel to wait it out or go on to a thunderstorm-clear area. Plan to arrive at the destination with lots of daylight remaining, three or four hours. Don't go on instruments with thunderstorms in the vicinity. Don't try to top them. Don't cut too close between two of them. Don't fly under the anvil overhang.

The progression of weather learning in thunderstorms is difficult, because any thunderstorms beyond air-mass ones will be frontal, and that may mean getting on instruments. Don't do this without radar and knowledge of how to use it.

We could fly under high-level warm-front thunderstorms

as a progressive step, but this is flirting with heavy rain and low stratus clouds that mean instrument flight.

After air-mass thunderstorms, therefore, the next step is a big one, because one may be forced to fly through a thunderstorm.

Flying thunderstorms in fronts and lows is the big time, and, unfortunately, has to be learned about the same way as swimming—by diving in over your head the first time. A person should at least know how to fly turbulence before he does it. Radar, as always, is a must. And this, again, isn't to help you fly through thunderstorms, but is just an extension of your tools for keeping out of them.

The learning steps we've outlined are not hard-and-fast rules; they are guides for one to start from and to think about. But there are a few firm points that should be rules.

1. Always have an out. Fly toward good weather or from good weather that you can return to.
2. Take on ice only after other experience has been gained.
3. Don't fool with thunderstorms for a long time.
4. Don't get on instruments near thunderstorms without radar.
5. Have lots of fuel.
6. Have lots of daylight remaining after the ETA.

There are other rules and they are in the book in many places, but if we summed up one weather rule it would always be: *Have an out! And know the big picture!*

Looking back over these steps, we can see that weather-flying experience isn't gained quickly. We need several seasons, years, to see the things we should see and experience the things we should experience. We must face the facts of weather flying. It cannot be gotten by injection, it cannot be gotten by reading a book, it cannot be gotten quickly. We must remain humble for a long time, and know when to quit or when not to go.

During the learning of weather flying we should be careful about reverting to VFR flying. To duck down low and try weaving in and out of mountains or even over flat terrain with minimum visibility is a sure way to trouble. Remember, VFR in marginal weather is the most dangerous way to fly.

It's important to realize that in weather judgment it often takes more courage to sit and wait than it does to go. The press of wanting to get somewhere will sometimes overshadow the gumption it takes to drag the luggage back to town and stay another night when you don't want to.

When I was a new copilot we were flying DC2s. I remember one of my first flights. The Captain was Jim Eischeid, a veteran with a special reputation for being an excellent weather pilot.

We were going to fly Flight 7 from Newark to Chicago. It was a rotten night with Chicago forecast low, ice en route. I stood next to Jim in the dispatch office, somewhat in awe, anxiously awaiting the flight to see how this man, whom no weather could stop, would handle it. Imagine my surprise when he turned to the dispatcher and said, "It's no good. I cancel!"

He knew when not to fly. Jim retired after a long career that went from open-cockpit mail planes to Connies, and he never scratched one of them.

Now we approach the era of all-weather flying. But will it truly be all-weather, will there never be a time when we cannot go? No, I don't think so. There always will be some time when it will be wise for a pilot to say, "I cancel!" The thing is to know when it's that time for each of us.

SUGGESTED READING

ANDERSON, BETTE RODA, *Weather in the West*. Palo Alto, California, American West Publishing Co., 1975.

EDINGER, JAMES G., *Watching for the Wind*. New York, Doubleday & Company, Inc., 1967.

DEPARTMENT OF COMMERCE, FEDERAL AVIATION AGENCY, *Aviation Weather*. Washington, D.C., U.S. Government Printing Office, 1965.

PETTERSSEN, SVERRE, *Introduction to Meteorology*, 3rd ed. New York, McGraw-Hill, Inc., 1968.

WALLINGTON, C. E., *Meteorology for Glider Pilots*. London, John Murray (Publishers), Ltd., 1961.

INDEX